Producer & International Distributor
eBookPro Publishing
www.ebook-pro.com

The Network of Time
Alon Halperin

Translation from the Hebrew by David Guerin

Contact: halperin.alon@gmail.com
ISBN

To Romy, Amit & Zohar

THE NETWORK OF TIME

ALON HALPERIN

CONTENTS

INTRODUCTION:

The Concept of Time – More Questions than Answers

Time – one of the most mysterious and fascinating riddles in the history of mankind, and also one of the most basic and useful concepts in everyday life. There is nothing that happens in our lives which isn't related to the concept of time; we think during every single moment in time, make decisions, arrive late, are in a hurry, wait, plan the future, reminisce, move forward and then go back. And yet, time might be the biggest riddle humanity has ever encountered, and one of the few subjects which have been discussed at great length and immeasurable depth for thousands of years without us arriving at even a single, uncontested insight. From ancient times to this day, both in western and eastern philosophy, giants of thought have tried to define time, discuss the riddles which surround it and determine whether it is real or merely an illusion.

How did it happen that human knowledge has developed so much over the years and today includes deep insights regarding the universe, humanity, biology, society and the processes which take place in these worlds of knowledge, without us being able to explain the most basic concept in this structure? Without us producing one single, simple and clear definition for it? Beyond the surprising fact that we can't manage to define and explain such a basic concept, this has far reaching implications for almost every field of understanding the universe and in our private world. Human knowledge was built on a foundation we do not understand, a foundation which, with time, could bring the whole structure falling down.

Listen to the ticking of the clock... now try to explain to yourselves: What is "time"?

As early as the 5th century, Saint Augustine wrote in his book "Confessions":

> "What is time? If no one asks me, I know;
> But if I wish to explain it to him who asks, I do not know."1

This is one of the most famous quotes regarding the concept of time and it appears in almost every text which deals with it, and is still thought provoking and surprising every time we read it. If so, how would you define this basic concept?

Intuitively we all understand what time is, however when we try to define this concept and give it a clear explanation, we hit a conceptual dead end. If you try to explain the concept of time to yourselves, eventually your definition will probably converge into two concepts which characterize our perception of time more than any other: "flow" (movement) and "change". Time flows, moves from one moment to another, from one event to another. A kind of flowing river of events. The second basic and intuitive concept of time is the concept of "change". Our first experience of time is the change we feel in the world. If this world would freeze like a picture on a wall you wouldn't ascribe an element of time to this picture. Time is the constant change we see, hear, feel and think. It's also the only expression we have for measuring time. In fact, the only ways we have for measuring time are expressed in some sort of change in the world: the movement of the clock's hands, the change of the display on the screen of a digital clock, the motion of sand flowing through a glass test tube, the movement of rays of sunlight across a stone slate and a uniform counting of the changing thoughts in our minds. You could never measure time itself without an external factor going through some kind of change. Every movement is measured by time but is actually measured based on another kind of change: velocity isn't distance divided by time but rather a change in location divided by the change in some clock according to which time is measured. Thinking about time without change leads us to a conceptual void. If the world would freeze from time to time and then get back on track without any change, there would be no meaning for the amount of time in which the world had remained frozen.

Notice how you haven't truly succeeded in defining what "time" is but only described your experience in time (the flow of events and a backdrop for the changes you see in the world). It would be like defining a lion using the words "fear" and "roar", or defining man using the words "love" and "hug" – it is a definition based on your **experience** of the object being defined rather than its true **essence**.

And yet, even if we accept flow and change as properties which describe "time" we run into paradoxes which will make it clear that what we define as the essence of time cannot be part of the definition of time. These two concepts, movement and change, include within them the concept of time to begin with. "Movement" is being located in one point in space at a given moment ("Time") and being located in another point in space at a later moment, meaning we define time using a concept ("Movement") which is itself defined by time. "Change" is the gap between an object's state in one point in time and another – that once again we define time using a concept ("Change") which includes the concept of time. These are of course circular definitions since we cannot define something using the same thing we're trying to define. We must therefore find a definition which relates to time's most basic essence without using the concept of time itself.

Don't take it personally – the problem doesn't lie in your analysis and defining abilities. No one has ever succeeded in giving "time" a clear and distinct definition which wasn't at least somewhat controversial, even though many have tried during the past 2,500 years. This is one of the reasons for the fact that there are those who claim that time is an illusion; if we cannot explain what "time" is, then maybe it doesn't exist.

The discussion of the concept of time leads us many years into the past, even before the times of St. Augustine. The most ancient discussion we know of which deals with the concept of time comes from ancient mythologies. Characters such as the Greek Chronos, the Roman Janus, the Indian Kāla and the African Ikenga appear in different mythologies in which time serves a role in the process of creation. The questions which stem from these mythologies discuss the concept of time in a vague way that's only hinted at, and mostly has to do with the creation of the world within time or the creation time as a result of the creation of the world.

The Jewish and Christian scriptures do not discuss philosophical questions regarding the concept of time, surprisingly, although it appears in later theological discussions in texts which are holy to those religions.

The truly interesting discussion began some time in the 5th century BC, almost at the same time in the west (In Greek Philosophy) and in the east (In Eastern Philosophy). The most in-depth discussion on the subject took place in Greece and focused on two central characters: Parmenides of Elea and Heraclitus of Ephesus. The two represent two contradictory perceptions of the concept of time, which are echoed to this day in modern discussions on time.

During the Middle Ages the discussion became almost entirely theological, and focused on the question of God's place and the creation of the world within the framework of time – did God create the world on the axis of time or was time created at the moment of creation? The concept of time served mostly to answer questions about the eternality of God and his omnipotence, and also to explain why the world was created at a particular moment rather than another. After the Protestant revolution a change took place in the discussion of time, thanks to the thought of Renè Descartes, Immanuel Kant and Gottfried Wilhelm Leibniz, among others, who tried to understand reality in a way that went beyond the limits of the Catholic church.

Only in the 17th century, mostly in Newton's work, did physics begin to touch on the concept of time and laid the foundation for the perception of time in the universe. This foundation would last for 250 years until Einstein arrived at the scene at the start of the 20th century and unify space and time into one entity and declared that time is not absolute but relative.

The great physical revolutions which took place during the 20th century as part of the theory of relativity and quantum mechanics which significantly affected the way in which we view time today and the language with which philosophy and physics try to deal with the ancient riddles from the times of Parmenides and Heraclitus. However, although the theories of relativity and quantum mechanics are seen as great revolutions in the physical perception of the world and the philosophy of the universe, in my view they do not constitute a true revolution in our ability to understand and define time. The concept of time as described by Newton and Einstein serves physics and those who apply it in engineering, medicine, space exploration, in the impressive technological applications of the recent century and in every other field as an efficient and practical tool, but this is not something which can appease the curiosity which wishes to understand the mysteries of the universe. The practicality of the current concept of time does not provide a clear answer to the philosophical discussion which attempts to explain the fundamental essence of the universe and existence. The bottom line is that in my view our ability to understand the big questions has not progressed significantly.

The staggering significance of this insight is surprising and provokes big, essential questions. Is it possible that over 2,500 years of development of human knowledge and revolutions in our understanding of the world – whether in philosophy, in physics, or in theology – our perception of time hasn't been significantly changed? Is it possible that the world views on which the 21st century understanding of the universe are based on such a fundamental concept which hasn't changed from the times of Parmenides and Heraclitus?

In this book I will recount the wonderful story of time, from ancient times up until the second decade of the 21st century; a wonderful story of a concept so basic and yet one that is magical and mysterious.

I will not limit myself to merely a historical description of the philosophy and physics of time, and will try to sow the seeds for the beginning of the next stage in the discussion. As part of this effort I will discuss the theory which came into our lives towards the end of the 20th century and which serves us today in almost every field of our lives – network and complexity theory. I truly believe that the insights which stem from this fascinating theory constitute a new and different world view which could provide groundbreaking explanations to old and complex questions, from understanding processes in our day to day lives to understanding fundamental concepts related to the entire universe, all in a simple and down to earth way.

In an attempt to unify the three great theories – the theory of relativity, quantum mechanics and network and complexity theory – I will try to describe a new view of the world, which includes within it the potential to better understand not just the concept of time, but any process in human life, in society and in our day to day lives.

Let us embark together on a journey in which we will encounter the great scientific and philosophical questions which occupy humanity in an attempt to understand reality – a journey in which we will face mysterious and fascinating riddles dealing with the universe in which we live, rethink everything that we take for granted, shatter all of the thought patterns we've been conditioned into and maybe even develop new ways of viewing the world, and new insights about the world and ourselves.

CHAPTER 1:

The Great Riddle – The Question of Time in Philosophy during Antiquity and the Middle Ages

In the Beginning, God Created the Heaven and the Earth

"In the beginning" are the first words appearing in the bible and they express a point in time. Although many commentators have discussed the meaning of the first words in the bible, the text has no philosophical, in-depth discussion of the concept of time. The theological discussion took place mostly during a later period, and primarily as part of the Christian thought of the middle ages.

The perception of the concept of time in the bible is summed up to a certain extent in Psalms 90:4: "A thousand years in your sight are like a day that has just gone by, or like a watch in the night." The bible presents two points of view on time: The first, the point of view of the regular axis of time which we experience as part of the beings of creation; and the other, the point of view of God who exists outside of time, as an eternal and infinite being which includes the unity of the past, the present and the future. The perception of time in this created world is identical to our intuitive perception of time and is expressed in an absolute and continuous axis of time which serves as a platform for the flow of events which take place one after the other. The deepest expression of this is the description of generations and sequences of events which are very common in the bible. God, however, has the status of observer outside of time and the created world, is not limited by time and perceives the past, the present and the future in one eternal and unified image. However, the bible includes many descriptions of God working within the axis of time, reacting to events, surprised by the behavior of his created beings, punishing, arguing and changing his mind. This twofold point of view, of God who sees the past, the present and the future from a single unified point of view, and yet acts within the axis of time, is the source of the known saying from Pirkei Avot "All is expected and permission is given". This saying presents several difficulties and questions, mostly when it comes to the issue of free choice which I will discuss at length later on in the book.

The biblical story of the creation of the world draws inspiration from Assyrian and Babylonian myths, and much like many early mythologies it views the act of creation as the source of time itself. This is why, according to the biblical story, the cause responsible for the creation of time must exist outside of time, in an eternal, timeless world. The question whether God created time, or whether the world was created as part of an eternal axis of time which had existed even before the creation of the world, will accompany the theological and philosophical discussion for many more years. The answer to this question can influence the perception of Divinity: Should God be seen as an eternal, supratemporal cause, or as an entity which also exists on the axis of time and is limited by it? More on this later.

The Flowing River and the Frozen Arrow – Heraclitus and Parmenides

An important early discussion of the concept of time began in the 5th century BC, both in Greece and in the far east. It was the beginning of a long journey of endless discussion, both oral and in writing, in an attempt to understand the riddle of time; a journey which has yet to be completed and one whose end we may never live to see.

The two famous pioneers of the philosophy of time were the Greek pre-Socratic philosophers Parmenides and Heraclitus.

Heraclitus of Ephesus was known as the "riddle man" or as "the dark one", due to his vague, laconic style. In short, precise saying he expressed his philosophical world view, which emphasized change as a central element of reality, and the unity of opposites. This view led to paradoxical sayings, which at times left Heraclitus' listeners more confused than they were at the outset.

His most famous saying, "Everything flows" (Panta Rhei), describes the understanding that movement and change are the basis of reality which is constantly changing.

On an intuitive level it is easy to relate to Heraclitus' approach. He describes our everyday experience of time: An axis or a flowing river of changing events, whose rate of movement we have no way of influencing. Another famous saying by Heraclitus adds even more depth to the concept of change: You cannot enter the same river twice. When man enters the river for the second time, it is no longer the same river, and man itself is no longer the same person he was when he first entered the river.

Even though this understanding of time is self-evident and matches both our intuition and everyday experience of time, human history teaches us that what is perceived by the senses does not necessarily relay the true essence of the world. Human intuitive insights have been disproved time and time again: from our view that lightning storms are an expression of the anger of the gods, to the idea that earth is flat and located at the center of the universe with the sun in orbit around it, to the scientific revolutions of the 20th century which completely contradict any intuition we ever had regarding the particles which the world is comprised of and the rules which govern the entire universe. Time after time human thought and science have managed to shatter our intuitive patterns of perception and shed a new and unexpected light on the reality around us.

But will the intuitive view according to which time flows from one event to another and serves as a substratum for the change we see in the world also be discredited and proven to be an illusion? In order to examine the problem, and possibly its solution as well, let us start with a contemporary of Heraclitus – Parmenides of Elea.

Parmenides claimed that change and movement are not possible since the world is one, eternal and frozen. Since change and movement are the fundamental expressions of our experience of time, then if change and movement do not exist then neither does time. Try and imagine a world with no change and no movement. Such a world is frozen; a static image which doesn't even include someone who's aware of its existence, since consciousness also flows along the axis of time, and if everything is frozen then there is no flow of consciousness and no awareness of the frozen world. In such a world it is clear that time has no meaning.

But what does Parmenides mean? How can one explain a perception which is so far removed from our intuition, which sees change and movement on a daily basis? How can we ascribe an illusory element to such strong and founded feelings?

It is very hard to explain Paremnides' true meaning based on his writings, since only several lines of his work survive - from the poem "On Nature" - and from those we can only understand his general view. But fortunately for us Parmenides had a diligent pupil, and his writings, in which he elaborates on and explains Parmenides' position, were preserved and passed to future generations.

The student, Zeno of Elea, wrote several famous paradoxes, beautiful and simple, which were meant to prove by negation our inability to rely on sensual perception. Thus, according to him, he has shown that the change and movement ascribed to the world are an illusion. And again, if there is no change and no movement - there is no time. These paradoxes remained unsolved for centuries, and to this day the solutions proposed are controversial.

I will not discuss all of the paradoxes, but I will mention the two central ones which are directly related to the question of time:

The paradox of Achilles and the tortoise - Achilles and the tortoise have a foot race. Achilles, who runs ten times as fast as the tortoise, gives him, of course, a ten-meter head start at the beginning of the race. The race begins, the ecstatic crowd calls out the names of the contestants, but it's clear to everyone present that the tortoise has no chance. We follow the progress of the race: since Achilles runs ten times as fast as the tortoise, while he's running the ten meters separating him and the tortoise, the tortoise had time to walk one meter. The race continues, and while Achilles runs the meter which separates him and the tortoise, the tortoise has had time to walk another ten centimeters. Achilles keeps on running and closes the ten-centimeter gap, but the tortoise maintains a distance of one centimeter between them. Achilles does not despair and runs the additional centimeter, but during this time the tortoise keeps going and walks another millimeter. Achilles feels that he can beat the tortoise and with the last of his strength runs the millimeter separating the two. The tortoise, to everyone's surprise, walks an additional tenth of a millimeter... And so, before the eyes of a mesmerized crowed, the race keeps on going forever. While Achilles closes the gap, the tortoise walks a tenth of the distance Achilles has covered - and Achilles will never overtake the tortoise!

While Achilles closes the gap between him and the tortoise, the tortoise creates additional distance between them. Since according to Zeno space and time can be divided ad infinitum, Achilles will never be able to traverse the infinite number of segments of distance and time between the tortoise and him.

What's going on here? In our experience of reality Achilles very quickly overtakes the tortoise. Why then, in Zeno's story, does Achilles chase the tortoise forever without being able to overtake him? How is it possible that Zeno's description, which sounds very logical, does not take place in reality? According to Zeno space and time can be divided ad infinitum, Achilles will never be able to traverse the infinite number of segments of distance and time between the tortoise and him. The fact that a rational and logical description of motion along an axis of time involves paradoxes and contradictions proves, according to Zeno, that in the true nature of the world, movement is paradoxical. That is, movement cannot take place the way we think it does and know intuitively. Therefore, what we experience with our senses is an illusion. And if movement and change are an illusion, then the continuity of time which is based on them is itself an illusion. Therefore, Zeno has succeeded in confusing us, if nothing else. He does however describe a real paradox which is created between our experience of the concept of motion and its rational description. However, does this help us understand what a world without motion or change really means? I imagine your answer is no.

Let us examine this argument further using another one of Zeno's paradoxes - the arrow paradox.

The Arrow Paradox - imagine that a Greek warrior fires an arrow at the skies of Athens. The arrow is moving through the sky and we're observing it. Now, choose a single moment in time and look at the arrow at this point. The arrow stands frozen in a space with no movement. At the next moment the arrow stands frozen in another point in space, static once again, with no movement. At any point in time the arrow is static in the air, at rest. If so, what is the source of the movement we see in its passage from one point to the next? The arrow does not extend itself in space in order to reach the next point in time, and if we examine what goes on between these two points in time in which it is motionless, we will find additional points in time between them where the arrow is once again at rest. So, if at any moment, ad infinitum, the arrow is at rest, how does it move and change its location? And if there is no change and no movement - as you already know - there is no time.

This argument isn't always easy to understand, and in fact it's one of the most powerful and difficult arguments to refute when it comes to the non-existence of time. In order to understand Zeno's argument, it's a good idea to use a modern analogy which conveys the central message which is relevant to our discussion. Think of reality as a movie filmed on photographic film. Anyone who's seen a filmstrip in his life (and the day when we'll only be able to see filmstrips in movies is not far off), knows that it is a collection of static images located in sequence on the film, and every image is slightly different than the one which came before it. Every moment in the movie is a completely frozen image. This collection of images does not include movement or change, and we experience the time progressing in the movie only when the images are shown in rapid sequence before our eyes. In the analogy of a movie film, our reality is in any given moment a single image frozen in time, in the filmstrip of the movie in which we're living; one image of the entire universe, with no movement or change. The next moment is different, but it too is frozen in time, with no movement and no change, and the same goes for every moment since the big bang to this day.

If so, the arrow in Zeno's paradox is static in every given moment in time, just as an image in a filmstrip is always static. Therefore, what we define as time consists of a collection of frozen images with no changes and no movement, and therefore with no time. Just like in the filmstrip, our reality has no change or movement either, but rather a collection of points frozen in time which our projector (our consciousness/senses) makes us experience as an illusion of continuity, change and movement along the axis of time. This is Zeno's meaning in the arrow paradox.

In these two paradoxes Zeno wishes to prove that in the real world there is no movement and no change; and since these are only an illusion, so is time. Therefore, we cannot define time since time does not exist and is an invention of our consciousness.

Since this claim goes against human intuition there have been many attempts at refuting it. And yet, I will surprise you by saying that this argument was never refuted or forgotten, and that in fact it has accompanied us for 2,500 years of heated debate, parallel to the development of human knowledge. To a great extent, as we shall see later on, it is also deeply rooted in 21st century physics.

In order to further complicate the problem and emphasize the gap between our intuition and experience and rational and logical thought regarding the concept of time, I will describe an additional basic contradiction in our ability to view time as "something" which flows. When Heraclitus argues that "everything flows" and describes time as a river, he sets a trap for us. Since the meaning of a flowing river is that at a given moment the river is at a certain location, and in another moment it is in a different one. In any moment in time the river is in a different location in space, and the way to examine its movement is to follow its movement along the axis of time. This is, of course, just a graphic description of the basic formula: Movement (speed) equals distance over time. That is, movement by its very definition includes the concept of time; there is no movement without time. Thus, we can easily examine at which speed the river is flowing, for example. However, if we attempt to define time using the concept of movement we arrive at a fundamental contradiction, and so the attempt to answer the question of the speed in which time is moving, for example, leads us to a dead end. This is because once we ascribe movement or flow to time, we're describing time using time. If the river is flowing along an axis of time and is located at any point in time in another point in space, then when time flows, at any point in time, it is in a different point in time. That is, we need an additional axis of time, along which time moves.

But this additional axis of time is also moving and flowing, and so in order to define its movement we need a third axis of time, and so on, ad infinitum. This is a logical contradiction of the infinite regress type: a contradiction which is created when a solution to a given problem recreates the problem it was supposed to solve, and the solution of this problem recreates the same problem and so on and so forth. This is exactly the contradiction we run into when we ascribe "movement" or "flow" to time.

And now we're once again back to the starting point, where our experience of time contradicts our perception of reality and logical thought. But by now we can understand, at least partially, the problem and philosophical and physical deliberation when it comes to the concept of time. Let us continue on our journey.

The Unity of Moments - The Perception of Time in the East

In an interesting coincidence, in the period where both Parmenides and Heraclitus were active, so did eastern philosophy - whose central currents were formed between the 10th and 5th centuries BC - begin dealing with the same question regarding existence and time, without the Greek and the Easterners having any influence on one another. All the currents of eastern philosophy (for example, Hinduism, Buddhism and Taoism) share the recognition of the unity of reality: the view that the internal relations between all objects and events, and even the experience of all phenomena in the world, are different expressions of one reality. This is an ultimate reality, inseparable, which is expressed in everything that makes up the world. The term in Hinduism which represents unity is the Brahman - the one eternal whole. The many gods described in Hinduistic mysticism and religion are too a different expression of the same unity. But for our purposes, the Brahman includes the unity of time and space.

The beautiful side of eastern philosophy is also its vague, confusing side - the inclusion of opposites: the ability to see the world through two contradictory viewpoints which come together to form one unified truth. The most famous expression of this is the Chinese Yin and Yang symbol, which expresses the fact that the universe consists of opposed but complementary forces. Therefore, in eastern philosophy, on one hand we see a connection to Heraclitus' saying that "everything flows", and on the other we also find all of existence in the present alone, with everything else being an illusion. Eastern philosophy treats the present as an eternal moment, which has no difference between past, present and future. And so, instead of a feeling of continuity between consecutive events - our everyday experience - we find one eternal, timeless and dynamic present, which folds the past and the future into itself. In this sense, both eastern and

Greek philosophy, as has been expressed in Parmenides' approach, views the flow of time as nothing more than an illusion.

However, unlike the west, where a contradiction is expressed in two opposing worldviews as we've seen with Parmenides and Heraclitus, in the east contradiction is the essence of the world. Therefore, in the eastern view the present is eternal, where the passage of time is an illusion, and at the same time dynamicity and flow are fundamental properties of that same reality.

We should add that eastern philosophy ascribes time and space to certain states of regular consciousness belonging to those who haven't experienced the transition to higher levels of consciousness. According to this philosophy, through meditation and leaving our regular state of consciousness, we find that conventional concepts of space and time are not the fundamental reality. Time is therefore just an expression of language and consciousness, before we've experienced the one true unity.

As we will describe later on in the theory of relativity, we see a unified concept of space-time which expresses a view according to which time is a fourth dimension, in addition to the three dimensions of space. This view is similar to that of eastern philosophy, especially that of Buddhism, according to which in the state of enlightenment space and time are interwoven without the possibility of separation.

The difference between the view of space and time in eastern philosophy and the way they are seen in western philosophy is fundamental. In western philosophy, as it has developed in Greece, there is a clear separation between the concepts of space and time and a view of the absoluteness of space and time as separate entities with an independent existence in reality. Space is the vessel which contains all the components of the universe, and time is the river which is constantly flowing in the background, and which is affected by nothing. This view is based on Euclidean geometry, which has influenced the understanding of the universe in Newton's theory. This theory had lasted for over 250 years, until Einstein came along and formulated a new one. The eastern view of space and time is more similar to the relativistic approach, according to which space and time are unified in a way that does not allow them the possibility of a separate existence in the world.

Between the two theories, Greek philosophy is more useful as a tool to study the world and as a practical way of action. It is certainly possible that for this reason useful science has developed especially in the west, between the middle ages up to the 20th century. Eastern philosophy is more reminiscent of the insights of physics in the 20th century, which view the world in both a larger scale (space exploration) and an especially smaller scale (the subatomic world); scales which, technologically speaking, we could only have been exposed to in the 20th century.

Eternal God - Theology in the Middle Ages

Did God create the world within the axis of time? And if so, is God limited by time? And what happened before he created the universe? From the biblical story of creation, difficult questions arrive for the believer, mostly due to use of the words "In the beginning". Many philosophers and commentators grappled with these questions, in Judaism and Christianity, and they influenced the perception of the concept of time in those religions and therefore on the perception of time of the entire western world.

The first theologian who delved into the concept of time and in its significance was Aurelius Augustinus, a Christian philosopher and theologian, a founding father of the church and one of the most influential thinkers in the history of Christianity. Augustine, who spent most of his adult life in Carthage and Italy in the 4th and 5th centuries, wrote an autobiographical book called "Confessions", and in it he dedicated an entire chapter to the discussion of the concept of time. Although one can find a good deal of influence by Greek philosophy, his writings include innovation and an in-depth, groundbreaking inquiry.

The perception of the temporality of the universe and the eternality of God is the basis to his writing, but he asks the difficult questions which follow from the story of creation in the book of Genesis and wishes to provide answers to them based on this view.

The first problem lies with the view that God created the world; assuming that time had existed before the act of creation, it is implied that God existed within time and was limited by it. That is, if time is eternal and God created the world at a certain point in time, then he is acting within the axis of time. If he acts within the axis of time, he's just as limited as any other being acting within time. As an answer to this Augustine claims that since time has no meaning in the eyes of God, he didn't create the world at a certain point in time but rather God **is** the moment in time. Therefore, from God's point of view, the world is constantly being created. But we cannot perceive this, of course, since we're part of the world which is limited by the framework of time. But then Augustine attempts to understand what the time we experience as beings of creation actually is.

And this is where Augustine takes time apart. Our movement from past to present to future is our way of experiencing time. If the past no longer has an existence, and the future doesn't have one yet, all that remains is the present. If we attempt to understand what the present moment is and break it down to smaller and smaller units, we find that the present moment has no duration and therefore it has no place in the world. The present moment is a meaningless point, with no content, since one can always keep on dividing it further ad infinitum. The present is a fiction. It is a transition between a nonexistent past and a nonexistent future. And therefore, time doesn't really exist. Time is the experience of transition from nonexistence to nonexistence, and so it is a product of the mind alone, and not an entity existing within reality.

Up until Augustine and the beginning of the middle ages, the philosophical discussion of the concept of time did not distinguish between two different views of time which they mixed together. The distinction between these two views, which will be described below, accompanies us since the middle ages to this day, and also serves in the long and complex discussion regarding the question of free will and free choice.

One approach is Presentism, from the word "present'. This is the view of time which recognizes only the existence of the present. According to this view, the past and the future have no existence, and our true experience is of the present alone - since only it exists in this world. That is, we're constantly moving from one present moment to another, and only it exists. The past no longer exists and the future has yet to happen.

The second view is eternalism, from the word "eternal". This is the view of time which recognizes a simultaneous existence of all times: past, present and future. Our experience of the world is our subjective point of view. However, everything we consider to be part of the past or the future has an identical existence to what we define as the present. The meaning of eternalism is that all the events in the past and the future have happened and already exist, and our experience shifts from event to event. It is similar to a geographical map where New York and London both exist, although one can't be in both these places at the same time. An event in my past and an event in my future exist simultaneously, I just can't experience them at the same time.

In order to better understand the entire discussion and complete our philosophical toolbox we need a distinction between two additional concepts. When we're talking about reality itself and the "things" which exist in it in practice, regardless of our perception, we're examining the world ontologically. Ontology (The theory of being) is our study of what exists in reality itself beyond our existence and our perception of reality. In contrast, when we study the way in which human consciousness perceives the world, and the limits of our knowledge, we're practicing epistemology (The theory of knowledge).

For example, if we're discussing the existence of the moon, we can make ontological statements regarding the existence of the moon and its properties in reality itself, before humanity was even created. We can also epistemologically discuss the way in which the moon is perceived by our senses and consciousness. In addition, we can discuss the gap between reality itself and the concept of "moon" which resides in our thought. This distinction between ontology and epistemology is fundamental in philosophy, and elucidates the questions we ask and the limits of the solutions we can produce.

We needed this lesson in the foundations of philosophy since these distinctions begin occupying an important place in the various views of time in the middle ages. While these distinctions are relevant to all the views and approaches which were common during the middle ages, in most cases these views and approaches included a mix of presentism and eternalism, without being aware that this mix has taken place. We will now examine the medieval approach through the prisms of epistemology and ontology, and through the lenses of presentism and eternalism.

The middle ages are perceived as the darkest period in the history of mankind; a period where human intelligence stopped in its tracks and a deep chasm of religious fanaticism, darkness, wickedness and ignorance opened up between the marvel of Greek creation and the renaissance. It is customary to define the middle ages as the stretch of time between the fall of the Roman empire in 476 AD and the discovery of America in 1492, in order to emphasize the divide between this period and the period which came before it, as well as the period which came after. However, although the middle ages were characterized by religious fanaticism, by instances of wicked brutality and a significant reduction of the limits of free thought, it is not a dark period when it comes to human knowledge. Some of the giants of human thought were active during this period and begot innovative theological and metaphysical thought. Christian theology flourished in this period and generated new insights regarding the world and our existence in it.

Two central characters in the world of Christian philosophy and theology in the middle ages were Anselm of Canterbury, who lived in the 11th century, and Thomas Aquinas, who lived in the 13th century. While most of their writing involved formulating and defining principles of Christianity and proving the existence of God, they also discussed metaphysical issues, including the question of time.

Anselm, when describing God, ascribes to him the perfect vision of all of time. Thus, he determines that the world is eternalist, that is, the past, the present and the future exist in it equally, and so God can see all times. As we've seen, this is a classical eternalist position which claims that ontologically (the theory of being) all times are things which exist in reality. Epistemologically (the theory of knowledge), even if we perceive this reality in our mind, in practice we cannot know all the different times all at once. We experience reality in a continuous, sequential manner, from one present moment to the next. God, however, is not limited in his knowledge of the world and so all times are accessible to him.

Thomas Aquinas, the successor to Anselm, was one of the greatest Christian theologians. He was declared to be a saint, and a portrait glorifying his image is presented in the famous Notre Dame cathedral in Paris.

Aquinas uses two metaphors when describing the relation between God and time: the first metaphor is that of an observer standing at the top of a tower watching a parade. From the top of the tower the observer can see the entire parade, however the spectators on the ground only see one part of the parade at a time, rather than all the people participating in the parade. This of course is an analogy to God watching the events of the past, the present and the future all at once, compared to the beings of creation who experience time only partially. The fact that the spectators don't see all the participants of the parade all at once doesn't mean they don't exist, which is why this metaphor supports eternalism - that is, the view that all points in time exist simultaneously, even though we experience them in a partial, sequential way.

Aquinas' second metaphor describes God as a point at the center of a circle whose circumference includes the various moments in time. In this way God is at the same distance from any point in time which exists eternally. This metaphor also expresses the eternalist view. But Aquinas doesn't discuss the eternalist implications of this view, and raises another claim according to which the only thing which exists is the present. Thus, he supports the presentist view.

A contradiction is therefore formed in Aquinas' ontological view. On one hand Aquinas describes an eternalist world where all moments in time have a simultaneous existence and are accessible to God, and on the other he claims that only the present truly exists. This contradiction, which appears in Aquinas' thought, will accompany us in the discussion of 21st century physics as well, especially when we discuss the implications of the theory of relativity and the concept of space-time.

So, as we've seen, the philosophies of east and west up until the end of the middle ages began touching on the deep questions regarding the concept of time, but they did so mostly from a theological and mystical point of view. In the next chapters a light of reason will shine on the world, and the reading in the book of nature, together with the reading of the holy scriptures, will lead mankind to new and fascinating insights regarding the universe and time.

CHAPTER 2:

New Horizons - The Early New Age and the New Philosophy

Pulling God out of the Machine - The Reformation and Descartes

In the morning of October 31st 1517 the priest Martin Luther left his home and hung the document of the 95 theses on the church door in Wittenberg, Germany, and thus began one of the greatest revolutions in the history of human knowledge. This document protested the Catholic custom of "Indulgence" (purchasing a note stating that God has absolved the owner of a sin that they've committed); however, by doing this Luther began the process of the reformation and founded the protestant movement, which changed the ways and tools through which we understand nature and the world.

As part of the European renaissance, the protestant movement placed man at the center of things and gave him power which he lacked in the middle ages: the power to understand the word of God, not through the mediation of the church but through personal reading of the holy scriptures.

In his book "Protestant Ethics and the Spirit of Capitalism", Max Weber links between the protestant revolution and the rise of private entrepreneurship, and as a result to the development of capitalism and the western economy. According to Weber, protestant ethics encourages action and hard work while amassing material wealth, which is perceived as a sign of a love of God and the nature of life in the next world. In addition, since God is the one who gives out opportunities and assets, legitimacy for economic and social inequality had developed. Scholars who followed him (such as Robert Merton) completed the historical explanation and linked this revolution with the rise of modern science as well. The understanding that knowledge can be accessible to the individual and that the individual must be proactive in order to achieve it was a significant perceptual change which had two important consequences relevant to our discussion: firstly, the urge to take responsibility and initiate led to an understanding that the individual can understand nature independently. Second, the insight arose that just as man can and should read the holy scriptures and understand them without any mediation, he can and should read the "book of nature" and understand it through direct observation.

As a result of these consequences, two approaches developed in the scientific revolution of the 16th and 17th centuries: empiricism (from the word empirical, experimental), which following the reformation called for direct reading of the book of nature, through observing nature and coming to conclusions based on what is perceived through the senses; and rationalism (from the word rational, logic), which was part of the counter-reaction against the reformation and called to understand nature through thought, without relying on the senses. The great significance of this process was that thought was no longer seen as the exclusive property of church officials, nor of the philosophers of nature, but as a product of initiative and personal responsibility. But this is not to say that this period was characterized by free thought or conceptual pluralism - far from it; many would be burned at the stake. However, at this point and as part of the protestant movement, understanding nature became a religious mission. Thus, began the process which would lead mankind through the scientific revolution, secularization and modernization, which allowed the development of science to this day.

Independent thought means doubting. Nothing in it is sacred: no pattern, no knowledge, no axiom. Everything is open to thought, contemplation and dispute. And what should one begin by doubting more than anything if not reality itself?

Seeds of doubt in the existence of reality have been sown as early as antiquity, in the east and the west. Gorgias, a Greek philosopher who lived in the 5th century BC, claimed that nothing truly exists, and even if reality did exist, human beings would not be able to perceive it. Plato too, in the famous allegory of the cave, attempted to prove the existence of an objective reality, although he claimed that it is very difficult to perceive and know it.

In the allegory of the cave mankind is likened to people sitting in a cave. They're looking at a side of the cave and see the shadows of those who pass through the entrance. They do not turn around towards the cave's entrance and accept the shadows on the wall as reality itself. At some point one of them turns around and looks at the entrance of the cave. The strong light coming from the opening is blinding and he has a hard time seeing the true reality outside. However, if he is patient enough and puts in enough effort the nature of the reality outside the cave will become clear to him, and the man will understand the hoax in which he had been living up to that point. The shadows on the side of the cave are the world of illusion, which we experience through our senses as reality, while the world outside the cave is the true reality which we fail to know and understand.

The allegory of the cave and the ontological and epistemological discussion have accompanied the west for centuries, but an in-depth discussion on the significance of the parable has waited for a time when a breach in the wall of religious rule appeared - during the reformation and more generally the early modern age. That's when a more meaningful process of skepticism began.

The first of the doubters would certainly be Descartes. In his book "Meditations on First Philosophy" written in 1641 Descartes writes:

"How often has it happened to me that in the night I dreamt that I found myself in this particular place, that I was dressed and seated near the fire, whilst in reality I was lying undressed in bed! At this moment it does indeed seem to me that it is with eyes awake that I am looking at this paper; [...] that it is deliberately and of set purpose that I extend my hand and perceive it; [...] But in thinking over this I remind myself that on many occasions I have in sleep been deceived by similar illusions! and in dwelling carefully on this reflection I see so manifestly that there are no certain indications by which we may clearly distinguish wakefulness from sleep that I am lost in astonishment."

Besides the fact that we learn that Descartes is sleeping naked in his bed (possibly the first written example of the phrase 'too much information'), this part also presents a clear example of rationalism, with someone doubting the senses and the existence of an external reality - since doubting the senses also means giving primacy to logical thought. And indeed, Descartes is considered as being the leader of the rationalist movement in western philosophy.

The question of the existence of reality will accompany us throughout this entire book and it is possible that it will never be satisfactorily solved. In the meantime it serves as fertile ground for cinematic pieces which challenge the concept of reality, whether in science-fiction movies such as "The Matrix" (1999), "Vanilla Sky" (2001) and "The Truman Show" (1998), or in classic movies such as "Blow Up" by Antonioni from 1966, "Rashomon" by Kurasawa from 1950 and many others. Time is an essential component of the discussion on the existence of reality, and the question arises: does an objective reality also include an objective entity called "time"? Perhaps time is nothing but an illusion, regardless of whether an external reality exists or not?

Descartes took it much further. In order to prove the existence of God and reality he wished to find the most fundamental insight possible, the most certain and primary truth which cannot be doubted and which cannot be ascribed to an illusion of the senses or a deception of a wicked demon. In a fascinating methodological maneuver Descartes doubts every thought which goes through his mind, until he arrives at the insight that thought itself cannot be doubted. Even if he's wrong, his mistaken thought still exists, and he exists as the one who thinks such a thought. This is where his famous conclusion comes from: "I think, therefore I am". This is the most certain basic truth.

Using the same method, Descartes breaks reality down to its most basic components in order to examine the question of the existence of its primary foundations. During this conceptual maneuver, of breaking reality down to its basic components, Descartes begins promoting a view which will become the most fundamental assumption in the history of science, according to which the universe is a "machine". If the universe is similar in its manner of operation to a machine, then all that is needed in order to know the universe is to break the machine down to its components and understand them. The process of breaking down and creating an explanation based on fundamental components is called reduction.

Descartes thought was greatly influenced by the invention of the mechanical clock, which although appeared as early as the middle ages, became widespread in every city in Europe during the 17th century. This invention was very significant and its influence lied mostly in two aspects:

1. Time - time became a very important factor in the life of the individual, in the economy and in the system of government. In every city a clock tower was erected, which dictated a clear and regular time that governed everyday activities. The fact that time had become a powerful factor in people's lives drove them to understand the way it works and contemplate its essence.

2. The universe - the mechanical clock presented to the people of that period a system made up of parts, the combination of which created a new and wondrous thing. The insight which developed from the wonder which was personified in the mechanical watch brought about an attempt to describe any system, including physical reality, as a machine. All it takes in order to understand the way the system works is to recognize the fact that the system consists of various parts, to take it apart, to understand the role of each part and which laws operate between the parts of the system. Through this method it would be possible to explain any system and the results it produces.

This mechanistic view of the world which sees every object in the universe - from the human body, through socio-economic system, to the solar system - a machine made up of various parts. In order to understand the system, one must perform reduction. Just as in order to understand the operation of a watch one must examine all its components - cogwheels, springs and dials - and the relations between them, so in order to understand a social system one must know its components (people) and the rules of interaction between them.

The mechanistic view, of which Descartes was one of the first and central proponents, deeply influenced the thought of Isaac

Newton and accompanied mankind for the next 300 years, and in a certain sense still does so to this day. And yet, as we shall see later on, this view also brought about the great crisis which accompanied the reception of the physical theories of the 20th century.

Imagine the people living during the 17th century sitting in the town square, looking at the large clock at the top of the tower, a complex machine working in a regular, uniform manner and which presents the fixed and periodic movement of mechanical parts, never stopping or deviating. Imagine this act of observing as a meditative act, which allows one to separate the fixed and circular movement which is taking place before one's eyes and time itself. Time becomes a separate entity, which is no longer only explained by the movement or change seen by observers, and it moves in the background with the observer having any way of influencing its speed, its rate of movement or its existence. When time becomes a fixed entity moving in the backdrop of the objects of our senses, it becomes a separate and absolute physical entity. Absolute in the sense that it has a separate existence, and the objects in the world cannot affect it.

And so, in a matter of a few centuries, we've transitioned from theological insights regarding the non-existence of time from reasoning related to the properties of God, to the perception of an absolute existence of time in the physical universe, which stems from an independent human way of looking at things which sees the world as a machine. The new view of time is the result of taking God out of the machine. And so, the pendulum of thought moves between the insights regarding the non-existence of time as a separate entity in the world, to intuitive insights according to which an independent time does exist and resides beyond the perception of our senses. This pendulum has been swinging constantly for 2,500 years, from antiquity and to our present time.

Descartes' concept of time is vague, sometimes contradicting itself, exactly because it is on the seam between the theological world view and the scientific, mechanistic view of the world. The contradiction between his views also reflects the approaches we described at the start of the chapter. On one hand, rationalism, which tries to understand the world through the power of thought and reason alone; and on the other hand, an empiricist view, which attempts to know reality through the senses and the intuitive insights of our everyday way of experiencing the world. In addition, with Descartes we can distinguish between the view of reality as a separate entity or as an illusion which exists in our consciousness alone, and the philosophical tools needed for the study of this subject: ontology wishes to understand what exists in reality, and epistemology follows the cognitive analysis of reality and the divide between our consciousness and external reality.

How then is the skepticism regarding the existence of reality reconciled with the view of reality of an absolute entity which is like a machine? How does the machine work on an axis of time which doesn't exist?

The conceptual and scientific revolution of the 16th and 17th centuries, represents, in my view, the fracture between philosophy and physics when it comes to the concept of time. Philosophy, with its rationalistic claims stemming from a logical and conceptual analysis, proves the non-existence of time. Physics, on the other hand, needs time as an entity with a separate existence in order to explain the phenomena which our senses perceive. This gap between philosophy and physics will find its clearest expressions in the comparison between Kant's and Leibnitz's philosophies on one hand, and Newton's physical theories on the other.

Consciousness and the External World - Immanuel Kant and the Study of the Brain

What is reality? Does it exist or is it merely an illusion? Do we experience the world as it is through our senses, which give us a reliable account of what is going on out there? Maybe we live inside the matrix and experience a sensory world view which is entirely different from what is taking place in reality itself? And perhaps we're even being directed and deceived by some sort of intelligent beings?

The answers to these questions are in the eye of the beholder. We were raised on the naive world view that there is a reality beyond our consciousness, that we perceive this reality as it is through our senses, and manage the sensory information (images, sounds, smells, touch, taste) which we received in our mind. Today we know that this simplistic world view isn't true. Throughout the generations several alternatives were proposed, and they can be classified using a wide spectrum of possibilities: on one end of the spectrum we assume the intuitive view which states that we experience external reality as it really is. On the other end of the spectrum is the matrix which we know, where we're characters in a computer game which projects "sensory" information for us, and reality itself is a completely different thing. And maybe there is no physical reality beyond our mind, and everything we experience through our senses is just an illusion, as the idealist philosopher George Berkeley claims.

Each one of us can place his world view somewhere on this spectrum. Do I believe that a physical reality exists beyond my consciousness? And if I do, just how much does my consciousness affect my perception of reality and alter it? The understanding that we're not a passive factor in the perception of reality, and that in some way we're perceiving and processing a reality in our senses which is different from the "real" reality which exists outside of us, has developed to a great degree thanks to the revolutionary philosophical work of Immanuel Kant. This understanding existed in some form even before Kant - in Plato's allegory of the cave it was made clear that something in our way examining reality causes us a certain form of blindness - and so we do not see the figures themselves but only their shadows. And yet, Kant says something else: we're built in a way that doesn't allow us to perceive reality itself, and it has nothing to do with the way we attempt to understand reality.

Kant was a German philosopher who lived in the 18th century and is considered as a central philosopher of the enlightenment age and one of the greatest philosophers in the history of mankind. His metaphysics and views of reality and law serve as some of the building blocks of modern western culture. He was known for his wit and humor, as well as his eccentricities. Due to his need to lead his life according to a particular and precise daily schedule, the citizens of his city would wind their clocks according to his regular daily activities. Even the church sexton would set the clock at the top of the tower according to the times at which Kant would take his walk, who himself would time his walks based on that same clock...

Kant was greatly influenced by the science of physics, and part of his philosophical view was created in an attempt to reconcile arguments regarding the question of time and space in Newton's theories (which we will discuss in the next chapter). Kant was also influenced by the ideas of Nicolaus Copernicus, a Polish astronomer who in 1543 published, in his book "On the Motion of Celestial Bodies", the conclusion that the earth is not at the center of the universe, or even at the center of the solar system. It is the sun which is located at the center of the star system and the earth revolves around it. It is hard to overestimate the significance of the transition from a geocentric view to a heliocentric one, which was later called "The Copernican Revolution". This was a revolution in the way we view the universe which shattered religious dogmas and which changed the understanding of mankind's true place in the universe. As a result, Kant would eventually perform what in time was known as the "Kantian Revolution". Kant himself even links between the two moves in one of his writings.

What is Kant's move? Firstly, he distinguishes between two parts of the world: the world of phenomena, which is the world as it is perceived in our mind, and "the thing in itself", which is reality itself, independent of our perception. The world of phenomena is the world which is perceived by our senses and which we experience, while "reality in itself" is not accessible to us and we can say nothing about it. These are two separate parts of reality but there is some connection between them. We cannot know "reality in itself", but it is the source of our sensory experience. Therefore, we cannot make any scientific claims about "reality in itself" and the field of science only studies the world of phenomena.

Kant's next move leads us to the concept of time and also explains why we don't have direct access to "reality in itself". Most of Kant's predecessors claimed that through our senses we receive information which is ordered in time and space regarding the external reality. When we experience a certain event, we receive sensory information regarding this event, which is linked to certain location in space and the axis of time. The source of information on the event lies in reality itself and we only perceive it through the senses. Or simply put, we see and hear exactly what goes on in front of us.

Kant reverses things and thus performs his "Copernican" move. According to him, the information we receive is not ordered. We impose in it a space-time order using our consciousness' interpretative filtering mechanisms. In other words, Kant claims that "reality in itself" provides us with unordered information, and it is our mind which filters the information and creates order and a division into categories in it. This means that the structure of space and time, the structure of causality and other structures do not exist in "reality in itself", but are rather filtering structures of perception, which only reside in our sensory mechanisms. That is, space and time, are not part of an objective external entity and do not exist in reality: these are structures which exist solely in our minds, and they allow us to link between objects and events in space and time. Time is our experience, and the interpretative structure of our consciousness imposes the dimension of time on our sensory experience. Time and space do not have an absolute existence in themselves but are rather ways of processing information.

The filtering tools in our mind are like rose-tinted glasses through which we see reality. The glasses actively affect the information which is received in our mind and make the world a slightly pinker place. Thus, the glasses of time and space order the unordered information we get from outside for us.

In recent years this approach is receiving interesting support from studies in the field of cognition, and especially from studies in brain research. The human brain is a complex organ and its abilities are surprising and awe-inspiring. Up until the middle of the 20th century, no significant study of the brain was possible, and the big questions were mostly discussed in the fields of psychology and philosophy. The question of consciousness, a riddle which itself has yet to be solved, and the question of how the brain perceives reality and how it analyzes the information it receives from the outside world, received a naive realist answer which described images of consciousness as mirror images of reality. However, recent discoveries in brain research present it as an extraordinary organ, and greatly change what we know about the perceptual and cognitive activity which takes place in our brain.

Naive realism, which to a great extent describes the common

view of the relation between reality and consciousness, claims that our brain has an "internal theater". This theater receives information from the external reality through the senses: sight, hearing, feeling, smell and taste. The information is projected on an internal screen in our brain, and the brain presents us with the picture or story which is revealed to us.

When we remember a certain image, we bring it up on the internal screen; when we're imagining we invent a new image on the internal screen. An illusion is an image which isn't directly related to an image which has an external source. This is our intuitive view of the brain and human perception. If I'm standing in the middle of a forest on a wintery day and I see green, wet trees swaying in the wind, I hear the sounds of the insects flying around me, smell the smell of rain and fresh vegetation and feel the mud which is clinging onto me and beginning to harden. According to the classical view, this information comes to me from outside, in accordance with its location in space and time, and it appears on the internal screen as an inclusive and unified image.

This might seem logical and even trivial, but this isn't what takes place in practice. Today it is known for certain that this description is not true. Firstly, on the most basic physical level, when we look at the world, the light which hits objects has its course shifted towards us and hits our retinas. We do not see the object but rather the particles of light which hit the object and which are "thrown" back to the eye's retina. We only see the messengers, which tell about the existence of the object in the reality in itself. As the philosopher Bertrand Russell: "To say that you see a star when you look at a point of light in the sky is like saying that you see New Zealand when you look at a New Zealander in London". This is also true when it comes to sound waves: do we hear the ambulance driving by us, or do the vibrations in the air set our eardrum in motion? Do we smell the cake, or do particles moving in the air stimulate the smell glands in our nose? Our link to reality isn't direct but is made through mediators which provide us with coded information regarding reality (mediators such as photons, vibrations in the air and chemicals).

On the level of cognitive and brain sciences, we know today that there are certain structures of perception of reality which exist in the mind in a genetic and inherent way. They're there from birth and precede the experiences and learning we go through in the world. One of these inborn structures, as follows from experiments conducted with day-old babies, is the expectation that an object moving before our eyes will maintain its shape, will not randomly disappear and be replaced by another object, and will traverse a continuous path without disappearing and re-appearing at some other random location. This view of the existence of objects in the world is inborn rather than learned, since we can find it in babies of the same ages all over the world, and it makes no difference what they've been exposed to up until then.

This is an example of the difference between our view of reality, which projects certain rules on the world which don't necessarily exist in it, and the world in itself (another discussion on this subject will take place later on in the chapter on quantum mechanics). It is possible that we impose our structures of perceiving reality on nature, in a way which is similar to what is described in Kant's epistemology.

Many experiments in cognitive science and brain research which were conducted in the last few decades empirically show that our experience of the world is a structure of our consciousness rather than reality itself. Time is also part of this structure, a pre-existing structure in our consciousness which actively processes the information we receive from outside. Time is our experience, and the interpretative structure of our consciousness imposes the dimension of time on our sensory experience. This does not prove that time does not exist in reality itself, and that time is merely an experience; however, there is no debate today regarding the fact that part of what we experience as time is a creation of our mind.

There are several beautiful examples which illustrate the difference between what we experience in our consciousness and the order of events which take place in reality.

One example is a famous experiment called The Cutaneous Rabbit, which was first carried out in Princeton by Frank Geldard and Carl Sherrick. In the experiment, light, rapid taps are made on three points on the arm: firstly, the back of the hand, then near the elbow, and finally, near the shoulder. In practice, the sensation felt by subjects are not a series of three rapid taps on three separate parts of the arm, but rather that of an animal hopping all along the arm in regular intervals, from the palm of the hand to the shoulder, and even in areas which were completely untouched. In other words, the brain receives sensory information of a certain touch and turns it into a story of consecutive touches, while it fills in the blank spaces between the points of contact in order to create continuity and uniformity. That is, the image of time discovered in the subject's consciousness does not match the image of time in the experiment conducted.

Auditory perception works in the exact same way. Sounds coming from one side of a stereophonic set are accompanied later on by sounds coming from the other side of the set, but in our auditory experience we experience a continuous transition of sound between the two sides. Even though the sounds come first from a speaker located on one side and then from a speaker located on the other side, our consciousness experiences a physical-like transition of sounds spreading across the entire room. Our experience is that the sound actually goes through the space of the room, even in the area between the two speakers, even though in actuality there are no speakers in that area. In the mind, a filling in of the gap between the two sides takes place. The mind creates one continuous and uniform path of sound in time and space, from one side of the room to the other.

The complexity of sensory perception and the way it is processed in the brain is also demonstrated in many optical illusions. But beyond the fact that these make nice stories, a fascinating insight arises here, one that I'm not sure you recognized on your first reading. The significance of these experiments is that our brain constructs an image of time which doesn't correspond to reality in itself.

Notice this point: if there are no sounds in the space between the speakers and if there are no taps in the area between the parts of the arm, how does our brain know that there will be such taps or sounds in order to fill in the gap before the next sound or tap arrives? We hear sounds from the right side and then from the left side, but before we recognize the sounds on the left side, we already recognize the sounds in the middle of the room. It is as though our brain predicts that an action will be carried out later in time, in another part of space, and in the meantime fills in the sounds in between. But our brain isn't a prophet, and the only explanation for this experience is that the brain builds an image of sounds and sensations only after they took place in practice and gives us a view of the world **in retrospect**. In other words, auditory information comes from outside, the brain analyzes the sounds coming from the right side and the left side, and instead of showing us this image it builds a completely different perceptual image: a continuous image in the room's time and space, which is what it displays on our internal screen. The brain "cooks up" a story about the information received, and only then brings the story up into our consciousness. There is no correspondence between the sequence of events in reality and the sequence of events of the experience appearing in our consciousness.

Another known effect of such phenomena is called the Phi Effect, and you can look it up on the web and find many examples of it. Points which are regularly spaced from one another flash in sequence create the feeling of a point moving along a continuous axis. And if they change colors in their flashing, the impression received is that the "moving" point is changing colors as it "moves" along its imagined path between points. Once again, the brain creates a continuous story and fills in gaps which don't exist in reality. Once the information is fully received the story is created in our minds, and in retrospect we're presented with continuous movement in time and space.

Those who are still having trouble accepting the great significance of these experiments and sees them as entertaining optical illusions and nothing more, will have trouble sticking to this position after reading the following experiment. One of the most famous experiments in brain research is called the Libet experiment, named after Benjamin Libet, a researcher from the University of California which conducted groundbreaking experiments in cognitive science in the 80s. In this experiment the subject is asked to perform a simple action (such as moving their wrist or one of their fingers) at random, according to his decision without any restrictions. The experiment examined two parameters: 1. The exact time in which the subject became aware of their desire to perform the action. 2. The exact time in which the neurological process related to the action being physically carried out began. In the Libet experiment and in many similar experiments carried out since then, something amazing was discovered: the conscious desire to move the hand came **after** the physical process or moving the hand in practice, in a gap ranging between half a second and several seconds. That is, the brain is already in a process of moving the hand before we're aware of our desire to move it. In other words, our desire comes after the action.

Beyond the fact that this is a reversal of what seems logical and intuitive when it comes to time, this experiment raises two significant and controversial insights. Firstly, once again we discover that the order of time, we experience stems from an illusion created in the process of perceiving reality, and that it doesn't represent a process which is continuous and consistent in time as described in naive realism. Time in reality in itself is not the absolute and continuous time we experience in our intuition.

The second insight is even more controversial and has to do with the question of free will. If our awareness of the desire to move our hand comes after the brain has already given the order to move the hand, then the desire didn't precede the action but rather the action preceded the desire. From this point we can conclude that our free will is merely an illusion. Some might claim that the fact that our awareness of desire comes after the desire in practice does not mean that the desire isn't free. However, the problem with this claim is that if we're not aware of our will, and if it does not stem from our awareness of choosing a certain thing, our will is not truly free. How can it be free if we're not even aware of it?

The question of free will is a deep and ancient question, much like the question of time. It has many points of contact with the question of time, and so it will also accompany us in the rest of our journey. Later on, we will examine it and its relation to the concept of time.

These phenomena suggest that the process of processing information in the brain is much more complex than we suspected. Time and the order in which things happen in our experience are not identical to the time and order of things in reality. We're still far from understanding how the brain perceives and processes the order of events experienced by the senses, but we already know that the brain isn't as innocent as we may have thought.

CHAPTER 3:

The Universe as a Machine - Newton and Classical Physics

And Newton Said... - The Principia Mathematica

The "Apple" company at the start of the 21st century is considered to be a symbol of technological innovation and human creativity. Its developments are the clearest example of the heights which technology has achieved and its extreme influence of everyday life of all of mankind - an influence which spans across continents, cultures, religions and languages. Technology and science in the 21st century are an expression of the most impressive human project to ever take place in the history of mankind - the scientific revolution. Even though mankind has gone through several significant revolutions on its way to the 21st century, never has such a significant revolution taken place over such a short period of time.

This revolution began about 330 years ago as a result of a (different) apple - the apple which fell on the head of Isaac Newton.

The story of the apple is probably just a legend, but great stars were always built and had their status established with legends and myths being built around them. And there's no doubt that Newton is one of the biggest of humanity's giants. There is a clear consensus regarding the status of Newton as one of the greatest physicists who have ever lived, a label which even Einstein (who was a public relations star in his own right) attached to Newton.

What made Isaac Newton, a reclusive, extremely religious, difficult and critical man, who rarely created social relationships and was even part of several public feuds with most of the intellectuals of his time, one of the most famous and respected people in the history of mankind?

It should be noted that beyond his scientific work, Newton mostly worked on religious philosophy, calculated the date of Christ's crucifixion (the 3rd of April, 33 AD, for those who simply must know) and tried to calculate the date of the Armageddon. Newton also dealt in alchemy and created compounds of various, strange substances. The religious component was also expressed in his scientific work. In his books he explained various phenomena in nature as long as he found mathematical support for them. However when he ran into phenomena which could not be explained as part of a mathematical framework, he made use of God to help him complete the picture (a phenomenon which was very common in descriptions of nature in those days and exists to this day in certain fields, but which became less and less common as scientific explanations went on improving).

Even with this assistance from God, Newton became the most famous and important scientist in the world following the publication of his book "The Mathematical Principles of the Philosophy of Nature" (Philosophiae Naturalis Principia Mathematica), or in short, the "Principia", in 1687 - the most important book in the history of science. The "Principia" was the first essay which supplied, scientifically, and with the use of a great deal of mathematics, an explanation for all the principles which operate the world and for all the phenomena we see in the heavens and on earth. This explanation provided answers for question regarding motion, space, time, the structure of the universe and the movement of the stars. These principles were received as the true version of the book of nature and served the entire human scientific enterprise for 250 years, until the scientific revolution of the 20th century. Another significant achievement was paving the way for the scientific method which combined empiricism and rationalism. Nowadays science is based on a combination of experimentation and observations of nature and insights which stem from thought alone (such as a definition of forces of nature which cannot be seen). This combination is the scientific method which Newton bequeathed to posterity (in the period after Newton, Kant unified empiricism and rationalism in his view of transcendental idealism, whose development was influenced by Newton's theory).

Newton defined three basic laws which describe the movement of bodies in nature, and thus created Newtonian mechanics which is a science which remains useful to this day. But we will spare ourselves this lesson in physics since it's not as relevant to our subject. The important thing is that Newton provided a basis for the world view which states that everything can be explained according to a few simple rules.

In the "Principia" Newton defines the concepts of space and time, he does this mostly because he uses them to create an explanation for the motion of bodies in nature. As early as his time it was clear that movement is relative to another body; that is, in order to define whether a body is in motion it must be stated relative to what the body is moving. When I'm in the belly of a ship moving in a constant velocity on the waters of a placid lake, I can't determine whether the ship is in motion. Even if I flip a coin it will fall at my feet, whether the ship is motion or not. The only way to determine whether I'm in motion is to look outside and examine the movement of the ship relative to another frame of reference, such as the mountains around the lake or the waters of the lake itself. Try and imagine an astronaut float in empty space with a constant velocity, surrounded by total darkness (a theoretical scenario where he doesn't see any stars). When he looks around him he doesn't see anything relative to which he can ascribe movement to himself. He doesn't feel any movement either, since there is no gravity which provides a feeling of resistance and there is no air around him which creates friction and a sensation of motion. An astronaut who has no way of feeling his own motion cannot subjectively determine whether he's moving or not. Physics also teaches us that if the universe is full of empty space (except for our astronaut), there is nothing objective which can be used to ascribe movement to the astronaut either. Motion can only take place relative to a frame of reference. Now, fill the empty space with water. Now we can know for certain if the astronaut is moving, since we can determine if he's moving relative to the water which surrounds him.

However, in space there is no water or any other substance filling up the vacuum, and Newton faces a challenge: how can one notice motion when there is no frame of reference against which motion can be defined? In order to solve this problem Newton defined space itself as an absolute frame of reference. In the "Principia" he characterizes absolute space as follows: "Space is absolute by its very nature, independent of anything external, always remaining equal and unmoving." In other words, absolute space is the container of the entire universe; we cannot sense it but it ontologically exists, as part of reality, and fills up the entire universe. Any body which moves in this container, moves relative to it. For Newton, when a planet is moving in space it's moving relative to other planets but also relative to space itself, even though we cannot see space, feel it or make empirical conclusions regarding its existence.

According to Newton, absolute space simply exists, period; even though he knew that no one has ever measured it. We can understand this view based on Newton's religious beliefs, since absolute space is in accordance with Christian theology, since God sees the entire world as a container which includes within it all the beings of creation. God is viewed as an external being which sees all. The description of a container which includes everything within it, with God above it, watching it, fits the point of view which the believer ascribes to God perfectly. Space then is one of God's "senses". In Newton's essay we can see a lack of satisfaction with his position, regarding the acceptance of the existence of something which we cannot see or measure and which we cannot influence. Newton therefore leaves us with a concept of motion based on a vague foundation, which isn't satisfactorily explained even by his own account.

When all is said and done, Newton's concept of space is intuitive and reflects the way in which humanity perceives space in everyday life. The experience of an absolute space, which serves as a constant frame of reference as a framework for motion, was created because of the random fact that we live on the planet Earth. The planet serves as a stable and fixed framework for the motion of objects. When we look at the sky, we see a relatively static image which served as a tool for navigation for many years. The fact that the night sky (beyond the solar system) is a fixed system lent support to Newton's definition of absolute space. The question now is what Newton has to say about the concept of time.

In the Principia Newton writes: "I do not define time, space, location and movement since they are clearly understood by all." Newton expressly states that he does not define the concept of time since it is understood by all, but this is a problematic statement which is far from meeting scientific standards. Even the many discussions and disputes regarding the mysterious questions which stem from the nature of time, which preceded Newton, suggest that the claim that the concept of time is understood by all simply isn't true.

Newton defines absolute time in very simple terms: "Absolute time, real and mathematical, flows of its own nature uniformly, with no regard to anything external." Newton's time is the naive, mathematical time, that of a river flowing in the background to which we have no access and which we have no ability of influencing, and it flows at a rate which is uniform, constant, and unchanging. Newton returns to the Heraclitus' concept of time and ascribes an objective existence, that of a physical entity in reality itself. Thus, Newton builds his theory on a shaky foundation, without any real justification beyond the fact that absolute time matches human intuition. Newton explains how time assists us as a background for the observations we make of bodies in the world, but it doesn't solve the basic question of the essence of time.

The big advantage of Newton's concept of time is how it functions as an excellent framework for studying the universe. Since if time is constant and moves uniformly all over the world, does not influence nature and isn't influenced by it, then it is a simply and convenient tool for making observations of nature and understanding the processes which take place in it. However, a relative and changing time impedes any possibility of making precise predictions and reduces our ability to watch nature and understand it. Thanks to this definition of the concept of time, Newton's science is very practical and served as fertile ground for the development of modern science.

Newton's theory had another great significance: it decisively provided a basis for Descartes' view which sees the universe as a machine. Newton took this view and turned the whole universe into a mechanical clock, which works according to absolute and known rules inside a closed "box" (absolute space), in a way that is constant and universal (absolute time).

Newton defined the fundamental laws of motion of every body in the world, the forces which work on those bodies and the interaction between the forces themselves. This system of laws and forces he placed inside an absolute space, fixed and clear, and he placed this whole setup on the axis of time, which flows in the background in a constant manner, without any surprises. With this world view he turned the entire universe into a machine, and all that is needed is to understand its components at any point in time, to apply the laws of motion and the forces which interact with each other to these components, and to observe their absolute state and the state of the entire machine at any given moment in space and time. Thus, he did away with the wondrous component of the world and made the universe completely predictable.

One of the most famous champions of the mechanistic view of the universe, and a follower of Newton, was Pierre-Simon Laplace, a French physicist, mathematician and astronomer who lived during the 18th and 19th centuries. Laplace published books which were based on Newton's theories, and which dealt, among other things, with the motion of the celestial bodies in the solar system. In his writings Laplace provided a basis for the mechanistic view of the world and developed it. He was considered to be the first person to ever describe the world in a way that was entirely deterministic, based on Newton's theory.

The most famous story about Laplace is the story of his meeting with Napoleon. Napoleon found out that Laplace wrote a whole book about the universe without so much as mentioning God. Napoleon, who liked asking provocative questions, asked Laplace in their meeting why God is not mentioned in his book. Laplace's reply was very bold for the time and has become legendary: "I had no need of that hypothesis". Napoleon was very much amused by Laplace's answer.

Laplace claimed that the present is a result of the past and the cause of the future, and as a result of this causal view he made the following claim: If there was an intelligent being (later this being was given the name "Laplace's Demon") large enough to contain the information on all the components of the universe, and which could also know where each of these components are presently located, then with Newton's laws of physics it could predict precisely what the universe will be like at any point in the future.

This is a radical view of the universe as a machine - since if we take a machine, examine every component at a certain point in time and understand the laws which define the relations between these components, we can predict the state in which the machine would be at any point we choose. If computers existed in Laplace's days he surely would have used an analogy of a supercomputer, which receives information regarding every particle in the universe, its location, velocity, and direction. And a powerful enough computer could apply Newton's absolute laws and could predict exactly where every particle will be at any future point in time. Thus, it would predict what would happen in the universe at any given moment.

This is, of course, an extreme deterministic world view, which assumes that Newton's laws completely describe the motion of the various components of the universe, and that every result has only one clear cause which leads to it. In a world where space does not influence the components of the universe and time moves in a uniform and constant manner, everything can be predicted in advance. And if everything can be predicted, man has no free choice.

The question of free will or free choice has accompanied mankind since the dawn of time and is even more ancient than the question of time. Laplace's view of absolute scientific determinism, which states that scientifically speaking we have no free will and that any choice is merely an illusion, is an extreme view. The great, unsolvable question of free will will accompany us along our journey, and eventually we will sum up the various claims and controversial conclusions in this matter.

The view of the universe as a machine is still deeply entrenched in the scientific community's view of the reality, even in the 21st century. Although this world view has allowed mankind to build its large reservoir of scientific and technological knowledge, it does not fully describe the processes we find in the universe, and so limited mankind's ability to break through significant obstacles in understand the processes of the physical, social and economic world, and perhaps in the world of the humanities as well. More on this in later chapters.

Absolute and Relative Time - Leibnitz and Clarke

One of the most bitter and personal arguments in the history of philosophy took place between Newton and Gottfried Wilhelm von Leibnitz, a German mathematician and philosopher, and Newton's contemporary. Unlike Newton, only those who've been exposed to the world of philosophy know his name, however, recognition of his work and insights has increased as time went by, especially in the 20th century. Today we can say that his claims regarding time and space match the current scientific and philosophical view much more closely than those of Newton.

The personal animosity which existed between these two people stemmed from different origins: national animosity between the continental German and the Englishman from the British Isles, animosity between two towering intellectuals who lived in the same times and the world was just too small to house them both, academic animosity stemming from the recognition each one of them received from the scientific and philosophical community and more. However, there were two specific subjects which increased the divide between the two. Firstly, and astoundingly, both of them invented a new mathematical field at the same time, which became one of the most practical scientific tools to this day: Newton, quickly followed by Leibnitz, had invented calculus. This branch of mathematics isn't relevant to our subject (even though it was mostly calculus which made it possible to solve Zeno's paradoxes), and so, for the purpose of you reading this book while fully conscious, I will forego the explanation. However, the argument between Newton and Leibnitz regarding which of them was the first to publish this new branch of mathematics turned the two into bitter enemies.

The second dispute between Newton and Leibnitz had to do with a philosophical and scientific world view, mostly related to the question of space and time. Leibnitz ruled out Newton's view of absolute space and time entirely and developed a relativistic view of these concepts. Leibnitz's solution and his dispute with Newton were documented in an exchange of letters between him and a student of Newton's, Samuel Clark, between the years 1715-1716. This dialog came to a premature end with the death of Leibnitz in 1716.

Newton claimed that space and time are entities which exist in and of themselves, without any dependence on the existence of objects and processes in the universe, and they are absolute since they are the vessels which serve as a means of communication between God and the world and stem from God himself (notice that in Hebrew the word "place" is one of the ways of referring to God). Leibnitz rejects Newton's view of absolute space and time and does not ascribe an independent existence to them: "Space and even time are nothing more than an order of possible entities which exist simultaneously when talking about space, or sequentially when talking about time, and they themselves have no reality."

According to Leibnitz, Newton's approach brings us back to the difficult questions of time's essence and time's progression. Newton does not discuss these questions since from a practical point of view his scientific method is compatible with the intuitive views of space and time. The problems raised by this view do not detract from its ability to provide accurate predictions and serve as an efficient tool for everyday scientific research.

But human history has already taught us that when we ignore difficult questions and try to build complex theoretical structures on a weak foundation, eventually the building comes crashing down. However, this crash didn't happen until the 20th century, when Newton couldn't so much as turn in his grave.

In a sense Leibnitz wanted to save science from these complex questions. It's hard to say that Leibnitz's claims are compatible with common sense or that they're indisputable, however we will describe them here in plain terms since I think they have a spark of insight which can assist us in understanding the revolution of the 20th century, and perhaps even clues to what is still to come, in my view, in the 21st century.

In Leibnitz's thought there are two important principles which are relevant to the concept of time, and it is those principles which lead us to conclusions regarding space and time. The first principle is called "The Principle of Sufficient Reason". This principle states that, simply put, everything has a reason. Everything should have a satisfactory explanation, and nothing can exist without a satisfactory reason. The second principle is called "The Identity of Indiscernibles", and stems from the first principle. Since everything should have an explanation, then if two things are completely identical, they must be one thing rather than two, since there is no satisfactory explanation for them being separate. For example, if two chairs are completely identical, located in the same place and the same time, the two chairs must be one and the same. The second principle sounds trivial, however this is a philosophical insight which can lead us to conclusions which are far from trivial. Leibnitz made use of these two principles in order to explain the concepts of space and time very differently from Newton.

When it comes to space, Leibnitz claims that if there is an absolute, uniform and infinite space, there is no meaning to the question why God chose to create the world in its current location in space, and not 10 meters to the right or at angle of 90 degrees relative to its current orientation. Since there is no satisfactory explanation for the location of the universe in absolute space, absolute space cannot exist. Leibnitz also makes a similar claim regarding time: if time is absolute, uniform and infinite, there is no satisfactory explanation for the fact that God chose to create the world in a certain point in time rather than a different one. Therefore, absolute time cannot exist. Similarly, using his second principle Leibnitz claims that if we cannot tell the difference between a world that was created in a specific location in absolute space and a world that was created somewhere else, it must be the same world. So absolute space has no real meaning, and if it has no meaning it does not exist. Furthermore, if we can't tell the difference between a world created in a certain point in time and a world created in another point in time, it must be the same world, and an absolute axis of time has no meaning.

To secular ears, these claims based on questions regarding God's ability to create a certain world rather than another one may sound strange. However, when we examine Leibnitz's claims while focusing on their rational and logical components, we see that these are claims with a philosophical depth that cannot be dismissed.

The counter argument Leibnitz makes in order to explain the phenomenon of space and time is even more interesting than the arguments which rule out the existence of absolute space and time. In order to understand Leibnitz's view of space and time we need to learn about another concept in his thought, called a "monad". A monad is a kind of fundamental component, an "atom" of reality. Everything we know is made out of monads. Unlike the material atom, the monad is an element which includes within it both the material and spiritual components of reality. In addition, everything that exists in reality is a monad, which also includes the relationships between it and the other monads as well. That is, Leibnitz builds an image of the world which is made up of fundamental components, where each such component - each monad - includes within it all its relationships between it and every other component. And so, space and time do not exist outside of monads but rather as a result of the relations between these monads. His view of time and space is relativistic. Leibnitz believed that the universe is made up by relations between the components of the world. These relationships are what define space and time, rather than the other way around. That is, time and space are not entities which house within them moving objects, but rather space and time are a result of the relationships between the objects in the universe.

In other words, space stems from the relative locations of objects. There is no vessel in itself called space in which objects move. If there were no objects, there would be no space; space is just the position of these bodies relative to one another. Time too does not exist in itself absolutely; the time we experience is just the location of events relative to other events, and without events, there would be no time.

In his fifth letter to Clark, Leibnitz presents an analogy which could make things more clear. He compares a regular tree in nature and a family tree. A family tree consists of branches which represent each member's relation to the other members (father, mother, children, cousins, etc.) The family tree would have no meaning without the relationships between its components. A regular tree exists in its own right as a tree, regardless of the relations between its branches, trunk and leaves; but a family tree does not exist without the relations between its components. Space and time are compared to a family tree; they do not exist in themselves but as a system of relationships between the objects of which they consist. In a way, it is like a network of relations, since the network itself does not exist without its relations and the relations are what create the concept of a network. We will link Leibnitz's view and the concept of the network when we discuss network theory, which began developing towards the end of the 20th century.

Regarding time and movement, Leibnitz makes claims about nature which send us back 2000 years on one hand, but also send us forward by 300 years on the other. In order to understand what Leibnitz said about movement and change we need to go back and recall one of the central problems regarding the concepts of time and movement. When we talked about Parmenides and Zeno, mostly in the context of the arrow paradox, we raised the problem of static moments of time and the riddle which still remains unsolved: where is the movement between these static moments? How do we see movement and change in a collection of static images which appear in the film of our lives in the universe?

This problem, which can be defined as the problem of the passage of time in a static image of time, accompanies us since the times of Parmenides and Zeno, to this day. Newton's theory strengthened the position of the static view, since if we look once again at time as a fixed axis moving uniformly, and we define the universe as a machine, then all we need to do is reduce it to moments in time to understand how the machine works. That is, in order to understand time we must dismantle it to its most basic components. We need to look at shorter and shorter moments, until we arrive at the lone moment that is the present, which cannot be divided further, and that is the moment of the static present. This act of reducing the system to its smallest components is exactly the reduction which characterizes the mechanistic view of the universe, which stems from Newton's theory. That is, Newton sent us back to the ancient problem of the question of the passage of time without offering a solution. As far as Newton is concerned, time just keeps going. He too had trouble accepting this assumption, but he eventually did as a result of a lack of choice and an understanding that its practicality trumps its philosophical and physical difficulties.

In the static view of time, movement is defined as change in space and time, and that's all the definition of movement consists of. This reduction of movement to locations in space and time is our intuitive and deep rooted view. According to this view, at a certain moment in time we're located in a specific location in space, and in the next moment we're somewhere else in space. This view can be relevant both in the framework of absolute space and time, as they were defined in Newton's theory, and in a relativistic system which views space and time as a phenomenon of relationships between components of the matter of the universe, as Leibnitz suggested in his writings. And so, the question of the passage of time isn't related just to the way space and time is defined; we need to examine what it is about movement itself, beyond its definition of an object changing position in space over time, which could be defined relative to other objects (according to the relativistic view) or relative to space itself (according to the absolute view). When we look at the world with a reductionist view, the same problem always arises; movement eludes us in the gaps between static moments and various positions, and we cannot find it.

Is there something else about movement which isn't included in the reductionist view? Something which will be added as a component of transition between moments and explain to us the riddle of the passage of time? What happens in between static positions? Is it possible that that objects in nature have an additional primary cause, which characterizes movement, and which can complete the definition of movement beyond changing position in space and time? That is, is it possible that objects themselves have an internal quality which explains the transition from one location to another?

At this point Leibnitz returns to views from antiquity, mostly those of Aristotle. Aristotle believed that matter has internal qualities and that movement and change are internal, active components of matter; they are not a result of an action performed on matter, which is the most basic component in nature, but rather they are fundamental components in nature and exist within matter itself. Therefore, they exist even in Zeno's static moments, and they are the source of the movement which takes places in the transition between moments. This view which Aristotle held, which breathes a kind of life into nature's basic matter, was not accepted as part of the scientific revolution, which, as a result of a desire to remove the wondrous component from nature, did away with matter's dynamic qualities.

Leibnitz represents the view that movement is something beyond change. The monad includes movement within it, that is, there's something about movement which is beyond change in position in space. Although change in position is part of what movement is, it is only one, insufficient aspect of it. In this view, the transition itself, the passage of time, is a fundamental and primary component of the universe, and doesn't just stem from the basic components of matter and position. Change is an internal quality of the object and not just an emergent property relative to other objects.

In order to visually illustrate Leibnitz's claim I will make use of a somewhat simplistic analogy; the distinction between a static image and a GIF file. When we take a video and take it apart to the moments on the "filmstrip" we get a set of sequential images, each one of which is completely devoid of motion, and is entirely static. But we know today of a type of file which isn't an image or a video, a GIF file is a kind of momentary image which includes motion in it; a kind of moment in time with a duration which includes a short motion repeating itself. Further on in our journey we will meet Henri Bergson and the "Duration" which he endowed every "moment" in time with, and we will try to make use of this analogy in order to draw some interesting insights regarding the construction of a new concept of "Time".

Put simply, Leibnitz solves the problem of the passage of time by providing the matter in the universe with another fundamental quality which exists even in the static moments themselves. This is a view which goes against human intuition. Intuition has a hard time understanding what a fundamental property of change is since it perceives movement and change as phenomena which can be observed when seeing a transition which isn't static. However, when trying to understand physical and philosophical claims which aren't intuitive, it's important to remember: nature doesn't owe human intuition anything, and the fact that man has experienced the world in a certain way in everyday life doesn't rule out any explanation which goes against this experience. We will need to remember this insight especially when talking about the physics of the 20th century, which is much further removed from human intuition than the claims made by Leibnitz. But this argument is a double-edged sword, since the problem of the passage of time has arisen mainly because, intuitively, we have trouble accepting the static view that says that time does not progress. And if we're to take our intuition with a grain of salt, perhaps we should have stopped at Zeno's claim and accept the fact that time does not progress.

For the continuation of our story it is important to remember these two views: Newton on one hand, with the machine made up of fundamental, static parts and which operates inside a closed box; and on the other hand Leibnitz, with the machine made up of fundamentals which have a certain "living" dynamic element and which operates in an open, infinite space.

We will return to Leibnitz and see that this view of change and time can be linked to advanced views from recent years. However, for now we can take a break from philosophy and get back to the "real" world. After trying to understand the passage of time, without arriving at any clear, unequivocal insights, we will try and answer another question: why does time flow from the past to the future, or to use a more familiar way of putting it: why does our universe have a distinct arrow, which points from the past to the future, and not the other way around?

CHAPTER 4:

From the Past to the Future, and Back - The Arrow of Time

Take any movie you know and play it backwards. At that moment it becomes a comedy. You've never seen, in the real world, a man jumping out of the water and landing on his feet at the edge of the pool. You've never seen a broken egg on the floor coming together and jumping back into the packaging, intact. You've never mixed your cup of coffee, and wondrously see the milk separating itself from the water and coffee beans to create three uniform, distinct layers. But why don't these things happen? Why does all this sound strange? Why is the scene where mixing the cup of coffee in just the right way in order to separate its ingredients is a fictional description which doesn't match real life? The answer to these questions may surprise you. The problem isn't in the laws of physics; this event could certainly happen physically speaking, it's just that the probability of it happening is extremely low.

In order to understand this answer and its significance in our everyday life, we should go back a bit and discuss the most familiar question regarding the concept of time: the mystery of the arrow of time. I'm glad to say that this question at least has a pretty good answer nowadays.

The mystery of the arrow of time is easy to explain: our way of experiencing the world goes one way. Most events we see in the universe seem irreversible, and any attempt to reverse them makes the event illogical. Why do all the events in the world have a clear time arrow from the past to the future, and why do we not witness any processes which would imply the reversal of the arrow? The initial answer would probably be: because it makes no sense. That is a possible answer, but in the world there are processes about which, simply saying they make no sense or go against intuition, does not constitute a convincing argument, scientifically speaking. The next answer is: it doesn't happen since it is physically impossible. Well, prepare to be surprised. In terms of the laws of physics, none of the processes I described are impossible. If we examine the math and the physical description of these events, we can reverse them in time and they would still match the predictions of math and physics. The reason is that the laws of physics are symmetrical in time - any such process is physically possible, in any direction in time. Here's a simpler example, whose reversibility in time might seem more logical: a white billiard ball moves on a surface, hits the black ball and sends it moving forward. Now play the film backwards: the black ball will move towards the white ball and set it in motion. This seems perfectly logical in both directions of time. In the same way, if we place a box of ice cubes on the kitchen counter, we'll see that they're slowly becoming small puddles of water. However, if the temperature in the room goes down drastically, the puddles of water will slowly freeze and go back to being perfect cubes of ice. These two scenarios might seem logical to us and we cannot tell the difference between different directions of time for these events. In the same way, the other physical processes are possible as well, but they don't seem logical to us since they go against our everyday experience.

If so, physics does not prevent these processes from happening, so why don't they happen? Why does the arrow of time go one way and events only seem to make sense in one direction of time, but not the reverse?

The answer to this question lies in the concept of temperature, or heat. If we understand why a body loses its heat in a cold environment and never grows warmer in such an environment, and why the heat of a cold body goes up in a hot environment until it arrives at an equilibrium with its environment, and why there is no spontaneous process of leaving equilibrium and going up or down in temperature - we'll understand the answer to the problem of the arrow of time.

Up until the mid-19th century, heat was explained with the concept of the "caloric". Caloric was defined as a viscous, invisible fluid which moves from place to place. Just as water tends to flow from high places to low ones, so the caloric tends to descend from areas of high temperature to areas of low temperature. During that century a new view was established, that claimed that the movement of the components of matter is what causes the phenomenon known as heat. And so thermodynamics was established as a branch of physics which best explains the phenomena of change in temperature. And indeed, one of the most fundamental laws in the world, with the biggest significance for our experience, is the second law of thermodynamics, which explains, for example, why we age and eventually die (pretty important, right?). The law claims that in every closed system, disorder can only increase. Disorder is defined using the concept of "Entropy". High entropy means a high level of disorder, and low entropy means a low level of disorder, or a relatively ordered system. According to the second law, entropy will always increase in any closed system. For example, at the start of winter, all the clothes will be folded in the closet, there is very little disorder and so entropy is low. With time the closet becomes messier and some of the clothes are scattered throughout the room;

now there is much disorder and the room's entropy is high. That is, when things are more scattered and disordered in a specific location, entropy is higher. Any room where an insufficient amount of energy is invested in tidying it, will move from a state of low entropy (a tidy room) to a state of high entropy (a messy room). This is the second law of thermodynamics and it applies to all the processes we know.

You might argue that there are systems that do not go from a state of order to disorder, but rather they become more ordered with time. For example, the human body becomes more and more ordered in its initial state of development. Does the development of the human body contradict the second law of thermodynamics? The answer is no; this body receives energy from a source outside of the system, and the second law of thermodynamics deals with closed systems. So in order for a system to create order from disorder, it must be open to external processes and receive energy from outside. In the case of the human body, the energy which comes from the sun and the food we eat allows the system to create order out of disorder. However, if we look at the human body, the sun and food as one closed system, that system too goes from low entropy to high entropy; the sun loses energy, and the creation of food and its digestion increase the disorder in the overall system. If we isolate the human body from its environment and look at it as a closed system, then due to the increase in entropy the body will go from being an ordered system to an orderless system, and eventually a dead system. In other words, every closed system must go from low entropy to high entropy. It is one of nature's most powerful laws.

The process of transition from order to disorder along the axis of time makes the second law of thermodynamics an exceptional one in physics, since it means that it is not symmetrical in time. Compared to the other laws of physics, which are reversible in time and do not lose their meaning when the direction of time is reversed, this law has a special significance since if we reverse the direction of time we lose the ability to describe reality as it is. This is why this law was investigated in an attempt to examine if it can explain the fact that the arrow of time only moves in one direction.

Statistical Time - Ludwig Boltzmann

The person who made the most significant contribution to the understanding of the arrow of time and building an explanation for this ancient riddle is Ludwig Boltzmann. Boltzmann was a tragic character in the history of modern science and struggled to prove his ideas for many years, faced with a lot of opposition from Europe's academic community. He put an end to his life at the start of the 20th century, just before his atomic theory, which had withstood many attacks, was proven beyond any doubt. His formula for finding the entropy of a system is engraved on his tombstone to this day.

Boltzmann examined the second law of thermodynamics and wished to explain why the arrow of time only moves in one direction, and why the level of entropy of closed systems in nature always goes up. Boltzmann's solution is genius not only because it explains the physical phenomenon we see, but also because it is presented with a great deal of simplicity. The greatness of many of the brilliant moves by the giants of science didn't stem just from the fact that a new, successful explanation was found, but also from the ability to formulate it clearly and simply.

Boltzmann based his explanation on the kinetic theory of gases. According to this theory gas consists of a huge number of particles which are in a state of constant motion. If we observe the gas found in a closed container, we see a huge collection of particles moving and colliding with each other. At the moment of collision particles transfer some of their motion between each other and change the direction of their movement. Just like billiard balls which move and collide with each other until eventually a more or less uniform distribution of the balls is created on the table, the same applies to particles of gas.

When gas particles are in a certain area of the container, they're in an ordered state (low entropy) and when the particles are scattered all over the container, they're in a less ordered state (high entropy). If we begin with a container in which the particles are ordered in a certain area, that is, the gas is in a state of low entropy, with time a process similar to the beginning of a game of billiard takes place: at first the balls are ordered in a certain part of the surface, and with time the ordered balls, which are in a state of low entropy, collide with one another and are scattered across the table, in a state of high entropy. Similarly, in the gas container the particles which are arranged in a certain part of the container collide with one another and "scatter" them, and gradually all the particles in the container spread out throughout the container.

Take for example an ice cube placed in a container full of air in room temperature. The air particles are warmer than the water particles, and when they're constantly moving and colliding with the ice cube they transfer some of their heat to the water particles. The cube begins to defrost, and so the water goes from a state of low entropy (centered in the ice cube in a particular part of the container) to a state of high entropy (a liquid gradually filling up the container). The air particles in the container, however, lose their heat, and the air temperature in the container will go down until it reaches equilibrium with the water's temperature. As long as we keep the container as a closed system (that is, the temperature outside of the container doesn't affect what goes on inside it), this equilibrium will be preserved until an external intervention takes place.

So in nature, processes go from an ordered state (low entropy), through motion and collision, to a more disordered state (high entropy).

However, why do we only see changes from ordered states to disordered states in closed systems? Why don't we see, every now and then, a cup of coffee which goes out of equilibrium and begins heating up, or a cup of coffee where mixing leads to a separation of the milk and coffee molecules, or a broken egg whose yolk and egg white come together and return to their initial state? Why is the arrow of time so pronounced? Why, as long as no energy is added to a closed system do we never see a reverse situation, which is perfectly possible according to the kinetic explanation we presented, since it makes sense that particles will collide with one another and every once in a while a situation will be created where the particles are more ordered than they started out as? Why does the collision between particles not create a state where, randomly, particles converge in a certain part of the container and reduce the system's entropy?

The explanation for this issue was presented by Boltzmann in a simple and awe-inspiring way, he developed the theory of statistical mechanics. Boltzmann claims that any state of the gas in the container is a certain arrangement of the particles in the container. There are arrangements in which the particles are all concentrated in one corner of the container, and there are arrangements where the particles are spread out uniformly throughout the container. There are different possible arrangements for the particles in the container, and we can define each arrangement as having low or high entropy. Arrangements where all the gas particles are concentrated in a certain part have low entropy, and arrangements where the particles are spread out throughout the container have high entropy. As we mentioned, there are many possible ways of arranging the particles in the container. Why then do we always see the options which have higher entropy? Only because the chances for these states are significantly higher than the chances of finding arrangements with low entropy. Take two extreme cases: in the first case the particles of gas are arranged in the shape of a pyramid in the corner of the container. In such a case there is a limited number of possible ways to arrange the particles. In the second case the particles are spread out throughout the entire container. In such a case the number of ways you can arrange the particles in the container is significantly higher than the number of arrangement options in the first case. Bodies in nature contain astronomical numbers of particles. The number of ways we have to arrange them is so big that it is impossible, in practice, to find gas in a very ordered state (say, in the shape of a pyramid in the corner), without an active effort to that purpose. This is why a closed system of particles goes to a state of high entropy, since the odds of that happening are infinitesimally close to 100%.

Another analogy which will help making the subject clearer is the Rubki's Cube. In a Rubik's Cube there is a huge amount of possibilities for disorder. To be precise, there are 43,252,003,274,489,856,000 such possibilities. However, there is only one arrangement where the cube is ordered. What are the chances that at random, after a few turns, we find the only correct arrangement of the cube? Next to zero. Now, we take an object which consists of an astronomical number of particles and try and think what the odds of them being arranged in a relatively ordered structure. The answer is zero!

This "fascinating" story about arranging particles in a container is important since it has an interesting implication for our everyday life. The fact that in everyday life we see processes in the particular way we see them stems from the same statistical principle, and from that principle alone. It isn't a law of nature. In other words, the fact that we always see milk being spilled on the floor, rather than milk going back into the bottle, is strictly a probabilistic matter. And so is the fact that water we've heated will become lukewarm, or the fact that we'll grow older until the day we die. The odds of the reverse happening, in each one of these cases, is extremely small, despite the fact that, theoretically, it is not impossible.

Back to the Beginning - The Hypothesis of the Past and the Multiverse

From the understanding that the arrow of time is merely statistical, an interesting question arises. If the probability to find matter in nature in certain arrangements is much higher than the probability of finding it in other arrangements, how is it possible that we always see matter in a state of low entropy, which is higher in the next point in time? If the probability of the actualization of certain arrangements is significantly higher, we should see them in a state of high entropy in the initial point in time as well, and in all past points in time. In other words, why is the universe in a state of low entropy, if the probability for that is so low? As we mentioned above, the universe is constantly moving from a more ordered state (low entropy) to a less ordered state (high entropy). If the probability of the universe being in a state of low entropy is next to nothing, then the state of the universe in the past is a state with near-zero probability.

The statistical explanation clarifies why processes go from one state to another, but it doesn't explain why we experience the arrow of time in one direction rather than another. Why does our universe have a transition from low entropy to high entropy, and why isn't entropy high to begin with? The year long discussion on the matter of why entropy keeps going up focuses on the wrong question. The understanding that thermodynamics is statistical proves that states of high entropy are the most common and natural ones, since the probability of them occurring is higher. Therefore, there is no need to ask why entropy keeps going up in the direction of the future rather than that of the past. Instead we should be asking: why are we in a state of low entropy to begin with?

The attempt to provide an answer to this question led to many different theories regarding the universe, and humans as a life form which evolved in a way which allows it to ask questions about the universe.

The simple answer to this question is based on the anthropic principle. We're not talking about entropy (entropia) here, but a similar sounding concept based on the Greek word anthropos: man. According to the anthropic principle, the universe we see has certain rules because only such a universe would make the development of life, and of intelligence, possible, and allow us to ask why the universe is the way it is rather something else. That is, if the universe was different in some parameter or other the development of intelligent life would not have been possible. For this reason, the answer to the question "why isn't the universe different" is that if the universe was different, we would not be here to ask the question. This principle is a problematic one, since it exempts us of keeping on inquiring and studying questions to which we don't have answers yet. However, in the case of the second law of thermodynamics, we can claim that the fact that the universe is in a state of relatively low entropy is what makes the development of life possible, and so the question why the universe is in such a state is irrelevant. The universe just is, this way and not another, and the person asking about it is a testimony of this fact. Even though this claim has its supporters, many wished to find the physical answer to this question.

Boltzmann himself was aware of the unique state of the universe and offered an answer of his own. According to him, it is certainly possible that the universe is, normally, in a state of high entropy and equilibrium which does not allow for the development of complex life forms. And since this is statistically possible, we may be living in a unique period in the life of the universe, where a certain fluctuation has caused the universe to exist in an unusual state of low entropy. The time we exist in is an exception in the history of the universe since the conditions created in it allowed us to evolve and discuss the current issue - as the anthropic principle explains. Boltzmann's suggestion is interesting mostly since he didn't know about relativity theory, quantum theory and the big bang theory. And indeed, the hypothesis regarding random fluctuations in the state of the universe can get significant support from quantum theory, which predicts the possibility of random quantum fluctuations in the universe (as a result of the uncertainty principle in quantum theory, which we will discuss in chapter 7). We can offer two criticisms of Boltzmann's proposal: firstly, we know today that the universe has existed for 14 billion years, and such a period of time does not allow - probabilistically speaking - such an extreme and exceptional fluctuation in the state of the universe. That is, the probability of such a fluctuation in such a (relatively) short period of time, in terms of the probabilities we're talking about, is almost zero. A more important criticism lies in the following question: if the universe is indeed a result of a random, rare fluctuation, why is this fluctuation so wide and complex? Carl Sagan, an astronomer and popularizer of science and science-fiction books, wrote in his book "Cosmos": "If you want to bake an apple pie, first you must create the universe." Sagan quite simply describes the problem in Boltzmann's argument. In order to bake a cake, we need its ingredients, and for that we need atomic elements, and earth, the solar system and the galaxies, the physical laws according to which the world operates, an oven, and so on.

If the fluctuation we're living in is random, the probability of it being so complex is negligible. It is much more logical that such a fluctuation would be created in a relatively small part of the universe, and in a very reduced way. It is much more likely that the only thing created by this fluctuation is the human brain, which holds all our information and memories of the world. For this reason, this argument is called a "Boltzmann Brain"; If we accept Boltzmann's argument, it is much more likely that what was created in this random fluctuation is floating brains in space, which hold our visual images of the universe which we call reality. In contrast, it is much less likely that the entire universe with all its galaxies, the solar system and Earth, with the chemical elements which enable life, and the single-cell organisms which will go on to evolve for billions of years into complex, intelligent life forms, sitting in front of a computer and grappling with questions regarding the concept of time... Such a development is extremely unlikely, and so it is very much possible that we're nothing but brains.

The most likely thing is that a random fluctuation will create a single brain, which includes all our information and memories of the world.

Even though this argument is very attractive, mostly for science-fiction literature, it doesn't help us when it comes to furthering our understanding of the universe, and so we will have to let it go for now in order to keep searching for a more convincing explanation.

David Albert, one of the most famous physicists of the end of the 20th century, coined the concept of "The Past Hypothesis" as an answer to the question why we're living in a world with low entropy. According to Albert, the only way to explain this situation is by hypothesizing that the universe begins with the big bang, for a reason still unknown to us, as a very ordered universe. That is, the big bang set the universe going with very low entropy, and from that point onwards it has been increasing. The big bang was the event which relativity theory predicted, in which the entire universe was so dense that it became a singular point, in which all the laws we know have collapsed. Relativity theory can't explain the state which existed in this point, and scientists hope that a theory will one day unite relativity theory and quantum theory (such a theory will be known as "quantum gravity theory") and will be able to explain the physical state of this point. We know today how to describe the various stages of the expansion of the universe, but we cannot explain the first moments of its creation, and so there is no satisfactory explanation for the fact that in the beginning the universe had such low entropy. "The Past Hypothesis" makes a lot of sense, and most physicists today believe it to be true, but we don't know why the universe was created in this format rather than another.

Even if we accept the assumption that the universe was created in a state of low entropy - and so has a distinct arrow of time, with irreversible processes aiming towards higher entropy - one question still remains open: why is our universe so unique and unlikely?

Another possible (and interesting) answer was raised by Thomas Gold, and is named "Gold Universe" after him. Gold says that if the universe is symmetrical in its physical laws, we must assume that the universe is symmetrical in its entropic processes as well. Just as at first the universe has low entropy, so it should be the case in the end. Gold assumes that the universe has two limits, a beginning and an end. At first the universe expanded and grew, and we're still in that stage of the universe and so we see the whole universe expanding at an increasing rate. However, at a certain state the universe will arrive at a point of maximum expansion and will then enter the great collapse, at the end of which it will return to being a singular point (and maybe then start back at the beginning again). According to Gold, in this process of the great collapse, the second law of thermodynamics will reverse its direction as well, reversing the arrow of time with it. At that stage in the future, all the processes in the universe will be in reverse to the ones we observe today and will move from high entropy to low entropy. Cups of coffee will become separated into milk, coffee and water; waves will converge back to their source; light will come out of telescopes and find its way in stars which will swallow it, and so on. Even though this description sounds very unlikely, physically speaking it is entirely possible. The "Gold Universe" hypothesis supposedly answers the question of entropy, since according to it the universe is symmetrical and isn't exceptional in the world of physical phenomena. However, Gold's explanation isn't compatible with the common assumption of our day, according to which the universe will not shrink in the future and will keep on expanding at an accelerating rate for ever. Today it is believed that the universe will not end up as ordered and singular but wide and in a state of equilibrium, a cold, "dead" universe, existing in a state of total disorder (aside from some light fluctuations here and there as quantum theory predicts).

Another theory which tries to explain the riddle of low entropy, and which is quickly gaining popularity in the physics community in recent years, is the multiverse theory. Generally speaking, this theory claims that our universe is only one of an infinite number of separate universes, which are being created each and every moment. The world consists of infinite separate "bubbles". Each bubble is a separate universe, in a cluster of universes developing without interacting with each other in any way. The place we live in is but one universe out of many. Notice that in the context of the arrow of time we're talking about a multiverse, and not the interpretation for quantum theory called the many worlds interpretation (which we will discuss in chapter 7).

How does the multiverse theory solve the problem of the arrow of time? If there are infinitely many universes, and each one of them is different in the parameters and laws according to which it was created and has developed, then it is very likely that at least one of them will be created in an ordered state of low entropy. The universe we live in is unusual, has low entropy and a distinct arrow of time, and it is different from an infinite number of parallel universes which have high entropy and a non-distinct arrow time. Sean Carroll adds a theory called "baby universes", according to which universes are being created all the time, due to quantum fluctuations, as bubbles which separate from out space-time and disconnect from it as separate universes. In this way an infinite process of new universes being created from the existing ones takes place.

The multiverse theory combines the theory which claims that there's an infinite number of universes in addition to our one, with the anthropic principle, which claims that we're living in this unusual universe because only this universe could support the development of intelligent life, and with "the past hypothesis", which claims that the universe in which we live was created in a state of low entropy.

Has the problem truly been solved?

Not really. There are two significant problems which make it hard for us to accept the multiverse theory. The problems are to a great extent philosophical, and so are difficult to discuss with physicists, who aren't always willing to accept philosophical arguments when the theory matches the math and the physical view. The first objection actually comes from Roger Penrose, one of the most esteemed physicists of the 20th century. Penrose calculated that the likelihood of a big bang which begins in a state of low entropy is 10 to the power of 10 to the power of 123. That is, from all the universes which have been created, according to the multiverse theory, the probability for our universe being created was extremely low. And if all the universes do indeed exist in tandem, then our universe, the one with low entropy, is high exceptional and marginal. Furthermore, Penrose claims that in order for life to be created it is enough that low entropy would exist in just a small part of the universe, such as our galaxy. But the universe has 100 billion galaxies, and according to the multiverse theory there are infinitely many additional universes. Penrose claims that this is a particularly wasteful explanation which does not match the basic criteria for an accepted explanation. The world of thought and science has long since accepted the principle which states that an explanation needs to be simple. According to the famous principle of "Occam's Razor", if there are two explanations for the same phenomenon and one is much more complex than the other, we should always prefer the simpler explanation. Penrose claims that such an explanation, which requires an infinite complexity of each universe and an infinite collection of universes, is highly unlikely.

The second objection comes from the philosophy of science. Karl Popper, a philosopher of science who was active in the 20th century, defined the "principle of falsification" which is accepted to this day by a large part of the scientific community. According to this principle a theory can only be considered scientific if it can, in principle, be falsified. The possibility of empirical support, without the possibility of falsifying the theory, does not testify to the theory being scientific. Take for example the theory that all swans are white. Even if we conducted 100 observations which showed that all swans are white, this theory would still not be considered scientific. What makes it scientific is the possibility that one day we will find a black swan - something which could disprove this theory, which states that all swans are white. That is, whether a theory is scientific or not is determined by the possibility of falsifying it. Another example raised by Popper himself and which is controversial to this day is the question whether Freudian psychology can be defined as a science. According to Popper, since we can explain any human behavior using the Freudian theory it cannot be falsified, and so it cannot be considered a scientific theory.

If the falsification principle is essential to determining what counts as a scientific theory, can we disprove the multiverse theory? This is a difficult question for which we don't have an answer yet.

At first glance it would seem that it cannot be disproven since we have no access to other universes, and so we cannot conduct an experiment which will confirm or disprove the question of their existence. If we can't disprove this hypothesis in any way, it should not be seen as a scientific hypothesis. There are researchers who believe that it is possible that the separate universes interact with each other when they collide with one another. If we find evidence for these collisions in our universe, we could empirically prove the existence of other universes. On in the future will we be able to know if the multiverse theory stands the test of reality. Nowadays this theory isn't falsifiable, and so according to Popper, it is doubtful whether it is indeed scientific. It should be noted that the validity of a scientific theory and the way in which scientific hypotheses and paradigms develop are complex and controversial subjects which are discussed to this day in the framework of the philosophy of science. The discussion isn't limited to the Popperian view and it has many objectors (we will discuss these questions in more detail in chapter 13).

At the start of the chapter I gladly claimed that at least when it comes to the arrow of time, physics provides a pretty good answer. Despite the lengthy discussion, I still claim that the arrow of time is one of the least troubling riddles we're facing. The statistical explanation for the arrow of time is an excellent explanation for the phenomena we experience and for the question why the world moves from low entropy to high entropy. The only question which hasn't satisfactorily been answered yet, in my opinion, is why we're living in a period in which the universe is characterized by low entropy. It's an interesting question, but in my view the central issue has already been understood. Furthermore, the past hypothesis is strong enough, and the discussion now shifts to the question of the state of the universe in the moment of the big bang.

Up until now we've dealt with the philosophical and physical questions raised by humanity's greatest minds up until the 20th century; we discussed the various issues and understood what's bothering us at the most fundamental level when we attempt to understand the concept of time. Now, that we're aware of the various explanations for these big questions, we will describe the great revolutions of the 20th century. As we will see, these revolutions completely changed what we thought about the world, and they clarify unequivocally that even if we think we discovered all the secrets of the universe, we will always be surprised by its ways and its complexity.

In 1894 Albert Michelson, a highly esteemed physicist, declared that most of the fundamental principles according to which the physical world operates have been proven. With this statement he expressed the general feeling which existed in the academic world in the end of the 19th century; science's work is almost done, and humanity is on the brink of a complete and total understanding of the entire universe. Lord Kelvin, one of the most famous physicists of the 19th century, which was involved in scientific breakthroughs in a variety of fields, gave a lecture in 1900 called "19th century clouds over the dynamical theory of heat and light". In this lecture he claimed that two lone clouds are obscuring the clear picture of the universe described by physics: issues related to the movement of light and the subject of black body radiation. Lord Kelvin, along with the rest of the scientific world, didn't know at the time that these two clouds will become the two biggest earthquakes in the history of science. From the question of the movement of light, the theory of relativity would emerge, and the question of black body radiation would lead to the quantum theory revolution. These two scientific theories will irrevocably change the way we understand reality as a whole, and time in particular.

CHAPTER 5:

Space-Time - The Special Theory of Relativity

In an international survey conducted in 2013, a list of the most famous people in the history of mankind was compiled. Albert Einstein was among the 20 most famous characters (alongside people such as Jesus, the prophet Mohammad and Julius Caesar). Very few are the people who do not know his name, but fewer still are those who can explain why he's so famous.

How did an eccentric Jew from Switzerland enter such a glamorous list? What makes Einstein so different from all the other important figures that appeared in the last few centuries? Unlike many scientists who changed our lives irrevocably, Einstein had almost no influence on our personal everyday life. Why then did he become the most famous scientist in the history of mankind?

The answer to this question has to do with public relations, timing and social and historical context, among other things, but the scientific answer was placed on the physics world's desk in 1905 ("Annus Mirabilis"). It was then that Einstein published, all at once, 4 articles which lay the foundations for a total change of everything which was known about light, time, space and matter in the universe at the time (or in other words, about pretty much everything, almost). In these four articles he proved Brownian motion, which established the existence of atoms and statistical mechanics, and the equivalence of matter and energy in the world's most famous equation, . He also proved that light is made of discrete portions which behave like particles, and thus paved the way for quantum theory which we will discuss at length later on in this book. However, Einstein became famous especially for his special theory of relativity (1905), and a decade later (1915) for the general theory of relativity which he presented to the world.

The fact that Lord Kelvin only mentioned two "clouds" which cast a shadow over physics' perfect and complete explanation of the universe, just 5 years earlier, emphasizes the tremendous revolution which Einstein's work was destined to bring about in modern science.

Vulcan and Ether – Something is rotten in the state of Newton

In order to understand this "revolution" we must go back to the end of the 19th century. Many scientists shared the feeling that after 2500 years of thought and research, the riddle of the universe itself was about to be solved. The Newtonian physics of the 17th century, which used absolute concepts of space and time, had still dominated science. Absolute time was defined as time which "moves" uniformly, without ever changing. Time "flows", and all the events in the universe take place relative to the same constant and uniform timeline. Absolute space is the frame of reference for motion in the universe. Recall the example of the astronaut floating in empty space surrounded by total darkness, without any ability of sensing his own motion. He cannot subjectively determine whether he's moving or not.

Newtonian physics clarified the concept of absolute space using the concept of the ether. The ether is the transparent substance contained in the universe (including empty space), and is an entity in its own right, to which objective and absolute motion can be ascribed. Space made out of a substance called "ether" is a clear expression of absolute space; space which exists in its own right as an entity independent of matter or movement. Relative space, on the other hand, is space which is defined by the relation between bodies and their movement and does not exist in its own right. The science of the late 19th century fully believed in the existence of the ether, even though this component is a product of human thought and has no observational basis. The move of adding physical entities to scientific theories in order to complete the picture and allow a satisfactory explanation isn't an unusual one and is entirely legitimate. Almost every new theory raises different claims about physical reality, some of which can be immediately verified through experiment and others remain, at first, as nothing more than a theoretical prediction. The aspiration being that in time science will develop and be able to empirically examine the theoretical claim and determine whether it gets accepted as a scientific fact. The question of what constitutes a "fact" is very complex and we will discuss it a little further on, however at this point it is important to remember Popper's principle of falsification, and say that hypothesizing the existence of a physical entity is legitimate so long as there is a way of disproving the theory of which it is a part. Until the theory is disproved, and so long as this entity matches the theory and the empirical findings, we can use it as a satisfactory scientific foundation. By hypothesizing the existence of the ether, scientists could go on and explain the universe based on the Newtonian theory.

The story of the French mathematician Urbain Le Verrier is a fascinating example of the chances and risks involved in publishing such theoretical assumptions. In 1846 Le Verrier claimed, following the work of several astronomers and mathematicians, that the significant deviations observed in the motion of the planet Uranus, compared to the mathematical predictions, should lead us to the conclusion that an additional planet exists nearby it, which affects the path of its orbit. The planet was discovered the same year by the German astronomer Johann Galle, and was called, as per Le Verrier's suggestion, "Neptune".

In 1859 Le Verrier attempted to solve the problem of Mercury's deviation from its predicated orbit around the sun. Following his success in predicting the existence of the planet Neptune, the deviation in Mercury's observed orbit compared to the mathematical prediction which was based on Newton's theory led Le Verrier to propose the existence of an additional planet. Le Verrier claimed that next to Mercury's orbit is another planet, affecting its motion, and he proposed to call it "Vulcan". Until his death in 1877, Le Verrier was convinced of this planet's existence, although it had never been observed. In 1915, with the publication of the general theory of relativity, Einstein's new equations explained Mercury's observed orbit without requiring the additional planet. Vulcan, unlike its friend Neptune, was forgotten in the pages of modern science's history.

What became of the concept of the "ether"?

Until the end of the 19th century, time was viewed as an external entity, moving uniformly and absolutely, without ever changing. This entity isn't affected by the processes in the universe and serves as a backdrop for observing changes in the universe and understanding them. Space was perceived as a transparent substance taking up the container we're located in, known as the universe, and any movement is defined relative to it.

In 1887 Albert Michelson and Edward Morley carried out an experiment which was meant to measure the velocity of the Earth relative to the ether. At the time, physicists were in agreement that light is made up of waves moving in the medium called ether, the way water waves move in a medium of water, and the way sound waves move in a medium of air. Without getting into the technical aspects of the experiment, the two scientists measured the speed of light in the direction of Earth's motion, and perpendicular to it. If the universe was full of a substance such as the ether, a difference in velocities would have been measured, in accordance with the prediction of Newton's theory. But the experiment was a complete failure: the difference found was negligible and well within the boundaries of equipment error. At first the results were interpreted as a mistake, and additional attempts were made in order to arrive at more precise results. Only 20 years later was the Michelson-Morley experiment recognized as the one which disproved the existence of the ether.

The central meaning of the results of the Michelson-Morley experiment was that the speed of light is constant for any observer and isn't relative to the Earth's movement. This finding contradicted the predictions of Newton's theory, and this is one of Kelvin's two "clouds" at the end of the 19th century.

However, anything in the universe can move in constant velocity. So what's so special about light's constant speed? And why did this question bother the physicists of the time so much?

The fact that light moves at a speed of about 300,000 kilometers a second (299,792.458 km/s to be precise), was already known at the time, however according to Newton's theory the velocity of any object in the universe is relative to a certain observer. If you're driving in a car at a speed of 100 km/h, moving towards a man who is standing in front of you, he sees you moving towards him at a speed of 100 km/h. If the same man is riding a motorcycle towards you at a speed of 50 km/h, he will see you moving towards him at a speed of 150 km/h, since the relative velocity is the sum of both of your velocities. If the motorcycle is moving away from your car at a speed of 50 km/h, while you're chasing him at a speed of 100 km/h, he will see you moving towards him at a speed of 50 km/h - the speed of the car minus the speed of the motorcycle. In other words, the velocity of an object relative to another is the result of the subtraction or addition of the two object's velocities.

The interesting result of the Michelson-Morley experiment was that light does not behave like this. According to Newton's theory the speed of light in the ether should be different, and it changes in accordance to the direction of Earth's movement as well. If I'm looking at a beam of light moving in the direction of Earth's movement, I should see it moving away from me at the speed of light minus Earth's speed. If I'm moving towards the light beam, I should see it moving towards me at the speed of light plus Earth's speed. Michelson and Morley showed that in any direction, regardless of the speed at which I'm moving, I will always see light moving in the same speed. That is, if I'm moving towards a flashlight shining its light on me at a speed of 300,000 km/s, I will see the light moving towards me at this speed, without the speed at which I'm moving towards it being taken into account. Even if I move towards it at a velocity of 290,000 km/s, I will still see the light moving towards me at a speed of 300,000 km/s. This phenomenon is unintuitive and goes against any kind of logic we're familiar with, and most of all it contradicts Newton's theory.

At the start of the 20th century, Albert Einstein, then a young man, was sitting in a Swiss patent office. Bored with his work as a clerk, he daydreamed about people chasing after a beam of light and thought about what a person moving away from a clock at the speed of light would see.

There is a historical argument regarding the degree of influence the Michelson-Morley experiment had on the development of Einstein's theory of relativity, although it is clear that he was aware of Maxwell's theory which stated that the speed of light is always constant. The contradiction between this conclusion and Newton's theory, according to which velocities can be added and subtracted, bothered Einstein greatly. He concluded that one of the theories must be mistaken. Einstein also loved thought experiments, and he often imagined unique situations in order to examine their physical meaning. As early as 16 he already imagined himself chasing a beam of light, a thought experiment which sowed the seeds of the theory of relativity he came up with 10 years later. This image, of a man chasing another man moving at the speed of light, which is constant, and even though he's accelerating he always sees that person he's chasing moving away from him at the same speed, will accompany Einstein in his understanding of space and time. From a different point of view, a man watching the chase from the sidelines sees how the chaser is increasing his velocity and can almost touch the person he's chasing. That is, the chaser sees the chased moving away from him at the speed of light since the speed of light is constant, and the observer sees two objects moving at near identical speeds, with one almost overtaking the other. Sounds strange? Welcome to 20th century science!

How is it possible that two observers can see the same event so differently?

Einstein understood that there must be some connection between the contradiction which is revealed in this strange image and Newton's view of absolute space and time. Einstein's understanding was also based on previous work, mostly on that of Ernst Mach. Mach suggested an idea which meant that absolute motion à la Newton was wrong. According to Mach, movement is relative to all the mass in the universe. Mach's space is like Leibnitz's space - it is a language which makes it possible to express the location of a certain object relative to another one. Therefore, a world with no objects renders the concept of space meaningless. This idea defangs absolute space and time and gives the matter existing in the universe the power of a frame of reference which is relative to itself. Mach didn't develop the laws of physics which are implied by this idea, and Einstein, who greatly respected Mach, did so himself.

Another image which would accompany Einstein was the thought of him riding a train and moving away from a clock placed at the station. If he does this at the speed of light, the hands of the clock will appear to be standing still since the light reflected from them will not overtake the train. But the clock in Einstein's hand will keep on working as usual, ticking at the normal rate. That's when Einstein came up with the idea which was to be the foundation of the special theory of relativity: Time can move in different velocities, in different places, in accordance with the velocity of your motion.

Everything in Life is Relative - The Contraction of Space and Time

Just as in any good story, in one spark of genius Einstein found the answer to the strange images which arose in his mind: what is constant and absolute is the speed of light, and what is changing is space and time.

Einstein realized that the speed of light is the highest velocity in nature, and so the speed of light is constant and absolute for anyone observing light, regardless of their own motion relative to it. If this is indeed true, it means that space and time are not uniform and absolute and that any object influences the progression of time and the contraction of space from its own point of view. This is the only possibility which doesn't create new paradoxes and which fits the mathematical description of the observations made in the world so far. That is, in order to solve the paradox of the chaser and the chased, the chaser's mind must be slowed down. Time slows the rate of its progression from the point of view of the pursuer, and so he sees the pursued moving away from him at the speed of light. Each one of them experiences time at a different rate than the other. Imagine that the pursuer gradually enters a state of "slow motion" and that the pursued keeps going at his normal rate. This image, of a pursuer moving in slow motion and the pursued moving at a regular rate, is seen by no one - since from the point of view of the pursuer time is progressing at its usual rate, and the same goes for the pursued. Although each one's time is moving at a different rate, they both perceive the rate of time's progression as they always do, since their brains are working at the same relative rate. However, physically speaking, the pursuer is slowing down but the pursued keeps on going as usual, and so from the pursuer's point of view, the pursued is moving away from him at great speed. An observer watching all this, however, who isn't in motion and who experiences time at a constant rate, sees a close race taking place between the two.

Since movement is composed of both time and distance, space itself must change in order for the image to be coherent. And so, in order to complete the picture Einstein claimed that space contracts as well, in accordance with the speed at which the observer is moving (each one of the participants - the pursuer, the pursued and the observer - experience space differently). He borrowed this idea from a concept called "Lorentz-Fitzgerland contraction", which was proposed to explain the failure of the Michelson-Morley experiment in regards to the ether. Lorentz and Fitzgerald proposed that space contracts as a result of the ether's influence, which is why the results of the experiment showed that the speed of light did not change. Lorentz was just one step away from the understanding of the special theory of relativity, but he didn't take the last step in time. Einstein took the idea and used it to promote the theory of relativity, thus decisively disproving the hypothesis of the "ether". One small step for Lorentz, one giant leap for Einstein's fame...

The faster a body moves, the slower time progresses for it compared to another object moving at a slower velocity. That is, Einstein did away with Newton's absolute space and time: there is no time moving uniformly and identically everywhere in the universe for all objects, but rather time is relative to the object's motion. There is no uniform and identical space everywhere in the universe for all objects, but rather space is relative to the object's motion. Only the contraction and dilation of space and time relative to an object can allow for the phenomenon of light's constant and absolute speed.

The beauty of this story is that it is not relevant just to light, but for any object in motion, at any speed. We don't feel this phenomenon in our everyday life only because we're moving very slowly and for very short distances compared to the universe. Even at velocities of hundreds of km/h, the influence on the space and time on Earth is negligible, and this is why we've been wrong for so long in thinking that Newton was right. Nature behaves in accordance with Einstein's description.

In October 1971, an atomic clock, which is an exceedingly accurate type of clock, was loaded onto a plane flying at high speeds at a great altitude. After the plane landed the atomic clock was compared to an identical clock which remained on the ground. Much to everyone's delight, the experiment confirmed Einstein's theory. In complete contradiction to common intuition, the clock on the plane showed that for it, time had passed more slowly than it did on the ground. That is, time on the plane did not progress at the same rate as time on the ground did and was relative to the movement of the plane. The theory of the relativity of time and the phenomenon of "time dilation" which follows from it have already been tested countless times since then and have been unequivocally proven. Particle accelerators demonstrate this phenomenon well. In installations such as CERN (the world's biggest particle accelerator, placed underground on the border between Switzerland and France) made it possible to accelerate particles to extremely high speeds, bordering on the speed of light, and to clearly see that the time it takes for them to decay is significantly increased. The fast particle experiences a slower rate of time, and so "lives" longer. This understanding is already being applied in our everyday life in the function of GPS systems. The driving instructions of apps such as Waze, for example, are received via satellite, which are adapted to a certain rate of time in space, and the different rate of time on Earth. Without the understanding of the theory of relativity, the instructions we get from GPS systems would get our location wrong, with error margins which go up to several kilometers.

While this goes against our intuition, we know today that the time we experience - the uniform and continuous axis of time flowing like a river as a backdrop to all the events of the world - is nothing but an illusion. This illusion stems from the fact that we're beings moving at low velocities and in areas with relatively weak gravity. Time in the universe is flexible and changes relative to every object. The velocity of an object and the gravity in its environment can drastically affect its personal rate of time. Our evolutionary development took place in an environment which doesn't require dealing with relative time in order to survive. For this reason, our perception of reality and the patterns with which we construct the order of events in our consciousness are very different from what is described in the theory of relativity. The neural structure of our consciousness and our perception does not allow us to understand relative and flexible time intuitively. In contrast, creatures which would have evolved in areas with very strong gravity, or creatures who could move at very high speeds, would need a very different way of perceiving the world in order to survive, and so would probably experience the world and time very differently from us.

Many make the mistake of thinking that Einstein unified space and time into one four-dimensional concept called "space-time, however this is a historical mistake. The first one to extract the unified concept of space-time was actually Hermann Minkowksi.

Minkowski was Einstein's teacher in university and wasn't particularly fond of him (at some point he even called Einstein a "lazy dog", probably as a result of an anti-Semitic view that was prevalent in European academia at the time). However, the tables turned and Minkowski was forced to accept Einstein's innovative theory. In 1907 he coined the term "space-time", which is one entity which includes both space and time. Minkowski treats space-time as a four-dimensional system: the three dimensions of space we're familiar with (length, width and height) and the dimension of time. An event is not located just in space, but rather it receives four coordinates - three coordinates of location is space and one coordinate of location in time. A certain event can have different coordinates in space-time for different observers. In a lecture he gave in 1908, Minkowski summarized for the first time the view of the unity of space and time as one entity, as opposed to the view that was prevalent until then, according to which these were entirely separate physical concepts: "The views of space and time which I wish to present to you today stem from the ground of experimental physics and that is where they draw their strength. They are radical. From this point onwards, space in itself and time in itself are doomed to disappear in the shadows and only a kind of unification of the two will maintain an independent reality."

Beyond the innovativeness of this statement, one can notice in Minkowski's phrasing the influences of eastern philosophy as I described it in the first chapter, and it echoes the concept of the unity of all of reality, including that of space and time.

"The Block Universe" - The Loss of Simultaneity in the Universe

The big question is what are implications which follow from the special theory of relativity? And the answer to this question is both fascinating and surprising.

The fact that every object has its own rate of time which is affected by the speed and direction of its motion completely rules out the concept of simultaneity in the world. What does this mean? According to Newton's view of absolute space and time, if we stop the universe at a certain moment, we can see all the events taking place in the universe at that moment. The universe appears as a frozen image of all the events in a given moment, and each such event happens simultaneously as all the others. While we're sitting on a couch reading a book, there are events taking place in the whole universe, and all these events are simultaneous to all the observers in the universe.

However, once we've come to know the special theory of relativity, we know that every observer has a different rate of time, in accordance with its speed and the direction of its movement. So two observers moving at different velocities will experience time, with the framework of the space-time dimensions, entirely differently. In this case, each one of them will have a completely different set of events which they will consider to be simultaneous. For each observer, a certain event will have different space-time coordinates, and so it will not be simultaneous for them.

There is no more "now", or an agreed upon moment which is the "now" of all the observers in the universe. Each observer has a different slice of the universe's events and hence a different "now". This is the most significant physical conclusion of the special theory of relativity. But this physical conclusion has a philosophical one which is far-reaching for the lives of human beings, and it's related to the concept of the “block universe”.

In order to understand the concept and this relativistic view of the “now”, we will use the example offered by Roger Penrose in his book “The Emperor’s New Mind”, which was based on the idea that had already existed in Minkowski’s time. Imagine that the people of the Andromeda galaxy (the galaxy closest to our Milky Way galaxy) are deliberating whether or not to go on a trip to Earth and attack its residents. At the same time, imagine two people on Earth, one of them standing still and the other moving at a very high speed, passing by him. When their eyes meet, we stop the universe. For them the encounter between them is a simultaneous event, and they have a common “now”, where they’re looking into each other’s eyes. But if we look at the faraway Andromeda, we will discover something amazing. For one of them, the “now” he and Andromeda have in common, that is, the simultaneous event which takes place on Andromeda the moment he looks into the eyes of his friend, is a meeting of the Andromedan military command, where the decision whether to attack Earth or not is supposed to be made. But in the eyes of the other one, the “now” he and Andromeda have in common, that is, the simultaneous event taking place at Andromeda when he looks into his friend’s eyes, is the departure of the attack spaceships heading towards Earth. That is, the event which takes place “now” in Andromeda is different for each one of them, and the moment they look into each other’s eyes, an event takes place for one of them which to the other counts as a future event. If for one of them the decision whether or not to attack Earth has not yet been made, and for the other the decision is already being executed, then the decision is certain even for someone for whom it hasn’t been made yet. If the event taking place in my “now” is an event which will happen in your future, then this event is a certainty for you as well. This view assumes a universe where all the events exist at the same time, a kind of four-dimensional block of space-time which includes all the events from the big bang until the end of days. A block whose every slice is a single point in time, a kind of "now", but every observer slices this block at a different angle and his slice includes different simultaneous events than that of his friend. In chapter 9 we will go deeper into the block universe argument.

All is Foreseen and Freedom of Choice is Not Granted - Absolute Determinism?

The immediate meaning of the block universe view is absolute determinism. Determinism is the philosophical view which states that all the events in the universe are pre-determined. There is a distinction between causal determinism and fatalistic determinism: according to the causal view, events are pre-determined because every event has a cause, and every cause has its own cause and so on. And so, every event in the past or the future will have an infinite chain of causes which necessarily lead to it. Unlike causal determinism, the fatalistic approach states that fate is what necessarily leads to a certain event, and no choice made will prevent the future from happening. Causal determinism however, looks at the causality which leads to a fixed future and not a future pre-determined as a result of fate. Determinism claims that we have no possibility of choice at all, and that any event, including our choices, is pre-determined. That is, if according to the "block universe" view all the events exist in tandem and each one of us experiences them differently in accordance with their speed, any event in the past or future have already taken place for some observer. And in that case, any event in which a choice is made in the future already exists, just like the case of the Andromedans in Penrose's example: the future of the observers is pre-determined, even if it hasn't happened yet in the "universal now" of each one of them, in the same slice of the block of events each one of them is currently in. The supporters of the "block universe" approach conclude that free will and free choice are just an illusion.

In order to understand their claim, we will use another analogy, which we used in order to explain eternalism. Suppose that all the events in the life of the universe exist on a map. When I look at a geographical map I know my location on the map, and when I search for the way from Paris to London I know for a fact that although I haven't arrived in London yet, London exists in reality and will be there to welcome me once I enter the city. The fact that I haven't reached London yet doesn't say anything about its existence in reality. The theory of relativity says something similar about time. The fact that I experience a certain moment of time as "now" demonstrates that I'm in a certain point in my "time map", but it says nothing about the existence and reality of any other moment in the past or the future on my map. In other words, even though I perceive myself as being in a certain point of my present, any other moment in my future has already happened and is waiting for me. Just like my entry into London will take place due to my movement in space (and doesn't indicate the non-existence of London before I arrive), my experiencing of a future event will take place due to my movement in time (and does not indicate the non-existence of this event before I experience it).

Einstein himself was a determinist and accepted the view that all the moments in our lives have an equal existence. The philosopher Karl Popper said that during a long conversation on this matter with Einstein, he called Einstein "Parmenides" due to this insistence on sticking to a four-dimensional universe with no difference between past, present and future. In a letter to the widow of Michele Beso, his close friend, Einstein wrote: "Now He has departed from this strange world a little ahead of me. That signifies nothing. For those of us who believe in physics, the distinction between past, present and future is only a stubbornly persistent illusion".

Even though the "block universe" is a strong argument, and is a conclusion directly following from the special theory of relativity which has already been proven, its deterministic implications are not perceived by most of us. The possibility of our lives being pre-determined and every future event of ours being out of our control is hard to accept. Every decision, every deliberation, every change in our lives, every successful or unfortunate event, is it all pre-determined? If so, how are we supposed to function in our everyday lives in light of this realization?

The question of determinism, which we will return to in the final chapter, has always been a great and controversial philosophical question. In the context of the "block universe" there have been different interpretations in order to deal with the deterministic implications of this claim.

The "block universe" is a physical reinforcement of the philosophical views we've encountered which claim that the universe is a collection of static moments (Parmenides), and it challenges the views which sees a dynamic nature in the universe with a component of becoming which is constantly developing (Heraclitus). Is there an interpretation for the theory of relativity which can avoid the deterministic conclusions of the "block universe"? On a philosophical level we can define the meaning of the future moment where the Andromedans arrive at a decision differently. The fact that the moment exists doesn't mean that the decision that's been made did not stem from free will. The future moment, from our perspective, where I make a decision exists simultaneously in the eyes of other observers since for them it already took place. But for me this moment hasn't taken place yet, and when it arrives I will make a decision based on free choice and free will. The difference is in the simultaneity of this moment for other observers, and not in the essence of the moment of choice, which is a moment in the developing continuum of moments on my axis of time.

However, this philosophical argument isn't enough to construct a world where there is no contradiction between physics and free choice. The solution involves an understanding of the physical theory. In order to do this, we will spend some more time on the theory of relativity. The meaning of the relativity of space and time is that for every observer in the universe there is, due to light's constant speed, a space with the ability to influence events and a possibility of being influenced by events. The constant speed of light limits our information and the ability of influencing certain events in the future. The space of events, called a "light cone", is any domain of events in space-time which light can arrive at from our present, to the future. That is, all the events within the "light cone" are events which we can see and affect, since they are within the range in which light can reach us in a given period of time, in the given speed of light which is known to us. If we compare the speed of light to the maximum speed at which information can travel, then the "light cone" is in essence the space of limited information each observer has. Anything outside of the range of the speed of light in the observer's space-time isn't inside their "light cone", that is, it's not in their space of maximum information and so isn't relevant to them. According to one claim, the meaning of the theory of relativity is indeed the existence of a four-dimensional space-time, however the simultaneity of various events in the past or the future lies in the light cones of different observers. And so, even if the events seem simultaneous to two observers moving relative to one another, the events are not simultaneous in time. The past will always precede the future, even if simultaneity is relative. In any case, it is impossible to transmit information regarding these events, since they're outside the range of information and are limited by the speed of light. These events shouldn't be ascribed any existence which goes beyond the regular axis of time of each observer, separately.

This view develops the theory of relativity to a different direction than the one at the start of our discussion: we are no longer talking about different events existing at the same time, and we experience their simultaneity differently. Rather the meaning of the theory of relativity is that every observer can only influence future events within their "light cone". According to this view, the theory of relativity does not discuss the ontological existence of all these events at the same time, and the deterministic problems which follow from it. Many people who deal with the subject hold this view, even if they don't phrase it in the same way. It is a complex idea which is easy to believe even without delving into its implications. Its big advantage is the ability to keep on functioning in the world without sinking in the chasm of determinism which rules out free will and the ability of free choice.

There are those who claim that these are musings which combine philosophy and physics and that the two can't truly be reconciled, since at the foundation of the argument, lies the concept of time, with all its confusing definitions and the divide between the "static-ness" of moments and our dynamic everyday experience of the becoming of new moments. The big question here is how to reconcile a four-dimensional universe which appears to be a static block of events, and the passage of time and its becoming. Later on I will offer a solution which combines quantum theory and complexity theory in order to take this discussion in a new direction, and in any case, I will develop the issue of the relation between the block universe and free will.

CHAPTER 6:

The Time Tunnel - Time Travel and Wormholes

Traveling through Time at 60 Seconds a Minute - Relativity and Time Machines

The year is 2135... Somewhere deep in the desert, in a classified location, the world's most secret facility is opened - the CAPF (The Center for Accelerating People to the Future). Lieutenant Colonel A. is placed in a closed cell which is located in a system of circular underground tunnels which go on for many kilometers. The closed cell is gradually accelerated together with lieutenant colonel A. inside the system of tunnels until it arrives at very high speeds, close to the speed of light. After lieutenant colonel moves for a while in the accelerator, he stops the cell, leaves the secret facility and discovers that he's arrived in the future.

Sounds like a scene from science-fiction movie? While we still don't know how and whether it's even possible to accelerate people to such velocities, it is only a technical problem. What is surprising about this story is that we know for a fact that physically speaking, if one day physicists and engineers succeed in solving this technical problem and build such a facility which can accelerate people to high speeds (similarly to the particle accelerator built in Switzerland), it will be the world's first time machine and the person being accelerated in this facility will be able to travel to the future.

Time travel is not only possible but in fact it happens regularly and takes place at every moment since the dawn of time. This journey is a journey we all go through every day, every hour and every second towards the next future moment. Our motion along the axis of time is a journey to the future.

Before you complain about my "silly" declaration, try to implement the insights of the special theory of relativity regarding our travelling in time and you'll find out that this is exactly what happens to us: each one's rate of time "ticks" in accordance with the speed of his motion. When we are standing still, our time progresses at a certain pace. The faster we go, our personal rate of time progresses slower relative to the other objects in the world. Imagine two people standing side by side, each one holding a clock. When they look at the clock together, they see that they're moving at the same rate: both moving at a rate of 60 seconds a minute, and every second is identical to the one that came before it, and so none of them can arrive at the future before the other. At some point one of them boards a spaceship and moves at a very high velocity relative to the other, who remains standing. When they look at their clock, they keep seeing that each one of them is progressing at a rate of 60 seconds a minute. But as we've learned from the special theory of relativity, the difference is that one second in the spaceship isn't equal to one second on the ground. Without them noticing, the second in the spaceship goes by slower than a second on the ground, and so when the spaceship comes to a stop and they meet again - they will discover that the time that went by on the clock which remained on the ground is longer than the one on the spaceship.

So, the faster we move, the faster we move towards the future, relative to our environment. Since if our time is slower than that of the environment, each second of ours equals several seconds of the environment, and in just a second we progressed several seconds towards the future. When we're moving at high speeds the environment's time moves faster, and the time that goes by is longer than the time which went by in our experience. For that reason, when we stop, we will find out that we've moved on to a future time. The only question is, at what speed are we moving. As we discussed, in the velocities we're moving on earth, the effect is negligible. As evidence, the world record holder for time travel, the Russian cosmonaut Gendai Padalka, moved at high speed in the space station above earth for a total of 879 days, and so when he returned to Earth he discovered that he leaped 0.02 seconds to the future (I hope that he managed to deal with the experience of future shock...).

This case is related to one of the most famous examples regarding travel to the future, which is called the "Twin Paradox" - although nowadays it is no longer a paradox and the question raised in this thought experiment have received a physical and logical answer. This thought experiment is based on things said by Einstein himself, without giving the phenomenon the title of a paradox, regarding two clocks. Later on, this developed into a thought experiment, relating to a pair of twins. The story is simple: one twin is sent to space at a high velocity, close to the speed of light, and leaves the other twin behind, on Earth. From the astronaut twin's perspective, the journey to space has lasted several years, and afterwards he returns expecting to meet his brother. But when he lands on Earth, he discovers to his astonishment that the time that went by on Earth is significantly longer than the time that went by on the spaceship, and now his twin brother is significantly older than him. The result of this thought experiment stems from the theory of relativity: the astronaut twin moves at a very high velocity, and so his rate of time has been significantly slowed down compared to the rate of time on

Earth. For this reason, the time that went by on the spaceship was shorter than the time that went by on Earth. The astronaut twin has arrived at Earth's future. This story is a clear example of the (albeit theoretical) possibility of travel to the future.

But before you rush off to build a time machine in order to travel to the future, you should take into account two "small" limitations: firstly, humanity doesn't have an aircraft that can get anywhere near the speed of light. The fastest aircraft ever built is the "Helios" probe, which was sent to the sun and arrived at a speed of 70 km a second, that is, 0.023% of the speed of light. Secondly, no living body could withstand the acceleration to such speeds and survive the way to space and back. But science never ceases to surprise, and various entrepreneurs have succeeded in offering technologies which could provide solutions for various problems. All we need now is for a startup to pop up which can solve the technical limitations.

So, travel to the future is physically possible. Is travel to the past similarly possible? Based on the physical knowledge we have today, the answer is no. Unlike travel to the future, and despite various attempts to find physical solutions for travel to the past, it seems that this isn't merely a technical problem, but that such travel isn't possible in terms of the laws of physics we know today. Another large problem related to travel to the past is the fact that such travel creates significant paradoxes, and so far we have no logical solution to them. The most famous paradox in this context is the "Grandfather Paradox". This paradox creates an infinite explanatory loop which cannot be untangled with logical and rational arguments. The paradox is as follows: a man chooses to go back in time and murder his grandfather. The grandfather's murder means preventing the possibility of the murderer being born, growing up and going back in time to murder his grandfather. So, the grandfather doesn't get murdered and he keeps the family line going, which leads to his grandson being born and to the moment his grandson going back in time in order to murder his grandfather... And so on and so forth. An infinite loop which cannot be solved. Another paradox related to travel to the past, called the "Bootstrap Paradox", was formulated in different ways over the years, and is mostly related to the concept of "information". Information is a quantitative entity which accumulates in a process which progresses in time, and so in nature we can't find information which as appeared ex nihilo. In order to understand the paradox created in travel to the past, we will make use of the following story based on a discussion Brian Greene conducts in his book "The Fabric of the Cosmos".

After we've gathered investors and overcome the technical limitations related to time travel, we've built the first time machine. Brian Greene, a physics professor at the University of Columbia, has been chosen to be the first time traveller. Greene arrives at 2100 and discovers that humanity as managed to create the "theory of everything" in physics. This theory manages to combine the theory of relativity and quantum theory and explains all the physical processes in the universe from the big bang to the end of the universe, including the question of time. With much excitement, he sits down to memorize the article published on the matter. Much to his surprise, the article is signed by Rita Greene, Brian Greene's mother, who never showed any interest in physics. He realizes that he must go back to the past, meet his mother again and teach her everything he knows about physics, so she can go on to develop the "theory of everything" later on and publish it.

When he returns, he tries to transmit the required knowledge to his mother, but quickly realizes that this is an impossible task given the time they have at their disposal. So, Greene decides to dictate the article he'd memorized to his mother. She writes it down, publishes it, and the rest is history.

Who then wrote the article? Who should receive the credit for the discovery of the physical theory of everything? Brian or Rita

Greene? Or perhaps a third party?

The most interesting question is, what is the source of the information which is contained in the scientific article the mother published? According to this story the information appeared supposedly ex nihilo, since the knowledge appeared in the article which Rita wrote after her son dictated to her the very words she herself wrote. And so, nobody wrote the article. This contradicts any physical process of accumulating knowledge. This paradox is unsolvable, unless we assume that it was created due to Brian returning to his mother in the past, and the paradox is assigned to the category of paradoxes created as a result of travel to the past. That is, time travel is possible only to the future, and not the past. And indeed, most researchers in the field will surely agree that time travel to the past isn't possible, even though there is no conclusive scientific proof as of yet.

There are, in my opinion, two main possibilities for avoiding these paradoxes. Firstly, we will recall that according to the "block universe" theory, the universe is a block of moments in time, a block of "nows" which include all the events in the universe at any moment, from the big bang up to the end of the universe. Every static moment has existence and reality, and any event in the four-dimensional universe is located according coordinates of space and time. We experience these moments in a certain order as a uniform, flowing continuum of moments. According to this view, even the moment in time in which I'm supposed to murder my grandfather has ontological reality, that is, any moment in the past exists just as it took place already and it cannot be changed. The conclusion is that I will not be able to travel to such a moment, and even if I do, I will experience it once again as the "me" who appears in that static moment. It is possible that I will be able to travel to certain moments only - in such a case there is a possibility of time travel, but only to moments in which I "existed". Since I didn't exist in the moment where I intended to murder my grandfather, since it took place before I was born, I cannot experience this moment in time. My only possibility is to go back to a moment where I had already existed and to realize this static moment without any change. This moment is simply realized again, in exactly the same way. In other words, time travel is just an additional realization of moments which have existence, without any possibility of changing them.

A second possibility of solving the paradox is through the many worlds theory. This theory was developed in the framework of quantum theory and I will discuss it at length in the next chapters. At this point it's important to know that it is an interpretation of the phenomena which follow from quantum theory, and its central idea is that an infinite amount of parallel moments are constantly being created in parallel universes. In every such universe the world is developing in a certain way. In one universe I write this sentence, and in another universe, I stop for a moment and take a sip of water. All the possibilities exist, but in separate universes. When I go back to the past and murder my grandfather, another split into two universes is created. In one universe I go back to the past, murder my grandfather, and end up not being born. But in my universe the murder has no effect on my existence, and I keep on existing without any memory of having travelled to the past.

The human desire to break past boundaries and to go beyond the limitations of time and space is a fascinating psychological subject which will always accompany us, but today it has no real significance. Which is why the discussion of time travel itself, which sparks the imaginations of many, is mostly possible in science fiction literature and cinema. It is possible that one day physics and technology will develop to a point where we can build time machines to the future or the past, or both, or it might never happen. But it's always best to keep dreaming, to go out and search for the scientific and technological loopholes needed to make such dreams come true.

Many precedents in human history have taught us that reality goes beyond anything we can imagine, and perseverance and persistence may break through the known patterns of the technological and scientific. However, we should always keep Hawking's question in mind - why don't we meet tourists from the future? Is it because a time machine will never be developed, or for some other reason? We will now examine a possible answer to these questions.

Time Worms - Time Travel through "Wormholes"

One of the physical solutions for travel to the past received a push from the world of science fiction in Carl Sagan's book "Contact" from 1985, which became a movie starring Jodie Foster. Sagan wanted to find a solution for travelling through time and space in the novel, and for that purpose used the term "black hole". He asked the physicist Kip Thorne to read the script and give him his opinion about it. Thorne had reservations about the idea of the black hole, claiming that the astronaut going through the black hole would surely die before he could go anywhere, but he suggested using the term "wormhole" as a solution for this form of travelling. The idea is based on the general theory of relativity. Versions of it have already been suggested at the start of the 20th century and were known as "Einstein-Rosen bridges" (ER Bridges), following the work of Albert Einstein and Nathan Rosen.

In order to understand what "wormholes" (a central concept nowadays) are, we need to go back to 1915 and discuss the second part of the theory of relativity which Einstein published: the general theory of relativity. The general theory of relativity deals mostly with the acceleration of objects and the concept of gravity, which weren't discussed in the special theory of relativity. Newton defined gravity between objects (according to the well-known legend of the apple which fell on his head), but he didn't define what this force was or at what speed it travelled. For Newton, gravity wasn't local, which means that its effect is immediate, without it having to travel through space. If Earth is attracted to the sun and the sun's gravity stops, at that same moment Earth will leave its orbit. Einstein, however, following the special theory of relativity, understood that the world is local and that everything has a maximum speed it can travel in, which is the speed of light. And so, gravity as well can only travel at a maximum speed, which is the speed of light. This means that even if the sun were to disappear, it would only affect Earth after 8 minutes - the minimal amount of time needed for light, or anything else, to reach the Earth from the sun. Einstein attempted to solve this contradiction between Newton's local view and the a-local one for a decade after he published the special theory of relativity.

Einstein's solution is amazing in its simplicity. According to the general theory of relativity, the sun doesn't exert any "force" on the Earth, but rather the Earth is in a constant state of falling towards the sun, and we and everything around us are in a constant state of falling towards the Earth. He borrowed this idea from another thought experiment (we've already established that this was Einstein's favorite hobby), which describes a man falling. Einstein realized that a man falling doesn't feel the force of gravity, and that the feeling of gravity only arises in us when we hit the ground or when we're going up in an elevator and feel the floor "pushing" us upwards. In other words, Einstein understood that "gravity" is a process of objects falling towards one another, and when we hit the ground or feel the floor of the elevator pushing us, we're feeling the resistance to our falling. This begs the question: why do objects fall in space? The answer is that space-time bends at certain places, and any object which is in a curved area, moves along the path created as a result of space's curvature.

A relatively simple way to understand the curvature of space-time which Einstein proposed is to think of a stretched sheet with an iron ball at its center. The iron ball causes the sheet to become curved around it, so that a depression is caused in the sheet. If we take smaller balls and place them on the sheet, they will roll towards the iron ball at the center, as if it were pulling them towards it. This is "gravity". In reality, however, nothing is being pulled towards the iron ball, but rather the objects are in a constant state of falling, and are being directed, along the sheet's curvature, towards the iron ball. In the same way, the Earth which is in a constant state of falling is directed along the space-time around it to fall towards the sun. The sheet is a two-dimensional space and is easy to imagine. In the universe, a body with mass causes a similar curvature in the four-dimensional space-time in its vicinity. That is, it bends both space and time, and affects the motion of objects which are close to it, and their rate of time. Earth, which is moving in its orbit around the sun is like a roulette ball going around the roulette's center. Relative to the center, it is exactly at the point in which it keeps orbiting (due to its falling towards the center) without leaving its path and getting too close or too far away from the center of rotation. So, Earth doesn't fall towards the sun and gets burned, and it doesn't escape its orbit and get swallowed up by empty space either.

If so, gravity is the falling of an object. We too are standing on the ground not because the Earth is pulling us towards it, but because we're in a constant state of falling and it is blocking our way. And since a large mass/energy curves space-time, a particularly large mass can create black holes whose curvature of the space-time around them is so big, that even light cannot escape it.

Initially, Einstein had trouble creating the mathematics of curved space-time, since he didn't have a suitable geometry. Like the rest of us, Einstein studied Euclidean geometry, which Euclid came up with in antiquity. As part of this geometry we deal with surfaces where the shortest distance between two points is a straight line.

Similar to an ant walking along a ball, thinking it's on a straight surface (but eventually returning to the same point she started in), so has mankind experienced the world (until one day it discovered that the surface isn't flat but rather round). The geometry of a curved surface is different, as attested by the parallel lines which cross the globe and meet at the north and south poles. Luckily for Einstein, his friend Marcel Grossman introduced non-Euclidean geometry to him, developed by Bernhard Riemann, a German mathematician from the 19th century. This geometry of curved surfaces helped Einstein complete the general theory of relativity. Without it, Einstein would have never breathed mathematical life into his relativistic philosophy.

One of the most interesting things which follow from the general theory of relativity is that in certain situations space-time can curve to a point where it creates connection between distant locations in space-time.

In order to simulate a two-dimensional surface, take a sheet of paper, fold it in half and pass a pencil through it, so a hole is created. An ant walking on the paper will have to walk along the page in order to get from one end to the other. However, if it discovers the hole that's been created, she would be able to traverse the page immediately. This is a "wormhole". Returning to four-dimensional space-time, a significant curving of space-time will cause a connection between distant locations. In such a state, the two parts of space-time, between which the journey would have been many light years long in the case of normal travel, could be connected by a "wormhole" which would allow for instant travel between them. However, how can such a "wormhole" be used as a time machine?

Theoretically, if we manage to build a "wormhole" on Earth we would have to leave one side of it on Earth in 2020 and send the other side to space at the speed of light. In such a case we will have a "wormhole" gateway in 2020 and another gateway accelerating through space, opening a time-gap between it and the gateway which remained on Earth. Travel through would immediately take us to the gateway which remained on Earth. This way we could build a time machine which would take us from the present back to a point in the past, and from the past back to the future. The time-gap created between the two gateways of the "wormhole" is exactly the twin effect we talked about at the start of this chapter. Except now we have two gateways of a time machine rather than two twins who can no longer recognize one another.

And now, back to reality...

Firstly, "wormholes" are theoretical mathematical entities which haven't been observed yet and without any way of implementing them in reality that we know about. Even if they do exist naturally (and this is contested), they probably exist only at the quantum level and in sizes comparable to the particle world, and so are not applicable as time machines which passengers could pass through. Even though many attempts have been made, a way of artificially creating macroscopic "wormholes" hasn't been found yet, and mankind is a long way from building such machines in practice.

Secondly, the math explicitly shows that "wormholes" are extremely unstable. That is, even if a "wormhole" is created, it will only be possible to preserve it for such a short period of time that even light won't have time to pass through it.

Thirdly, travel will only be possible to the exact time at which the "wormhole" was created, and so, as long as such a time machine hasn't been built yet, it will never be possible to go back to an earlier time. In other words, even if in 2025 we succeed in building such a time machine, we will only be able to go back in time up to 2025, since the gateway remains in the same place it was when it was first opened. This could also be the answer to Stephen Hawking's question regarding the absence of tourists from the future. Even if travel to the past takes place one day, it won't be to our past, but to a future point in time in which the machine is first built. This is why we don't meet visitors from the future, and why we won't be meeting them any time soon.

Despite the publication of the special and general theories of relativity, Einstein received the Nobel prize in physics in 1921 for his 1905 article which discussed the photoelectric effect, according to which electromagnetic radiation which hits a metal causes the emission of electrons. The explanation Einstein provided for this effect reinforced the understanding of the particle nature of light and the development of the term "photon". This effect, which is one of the most important foundations of quantum theory, reinforced the understanding that the universe is even more mysterious than we thought. We will discuss this theory in the next chapter.

CHAPTER 7:

The End of the Age of Certainty - Quantum Theory and the Measurement Problem

Billiard Balls and Water Waves - The Wave-Particle Duality

Quantum theory is magic.

When we observe the actions of a magician, we realize that he's showing us a beautiful illusion, which arouses wonder in us, specifically around the question: how does he do it? Any sensible person understands that acts of magic are an illusion. When the magician presents us with the "magic" trick, he leaves out information that's critical for understanding the surprising result, eliciting applause from the audience. It's clear to us that if the magician had only shared all the information regarding the process with us, the surprise and wonder would have vanished; at the end of the day every phenomenon in nature has an explanation that's clear, consistent, continuous and complete. The scientist in us will search for the physical, chemical, biological, etc. reasons, and the believer in us might be satisfied with spiritual, theological or mystical reasons, stemming from a higher power, which is entirely beyond our understanding. However, the explanation will always be complete and continuous.

Quantum theory is an example of magic which isn't an illusion.

Quantum theory describes processes which cause surprise and bewilderment and is sometimes even received with a thunderous applause. However, no sleight of hand or concealed information is involved. Today we can say with a good degree of certainty (insofar as science provides certainty) that this is how our world truly works, and that this way goes against anything we're familiar with when it comes to the reality around us. Everything we were raised on, everything we were taught about the world - the certainty in the world, the causality, the lack of randomness, the solidity of matter, the ability of objects to influence one another, the ability to divide the world into various, distinct categories - all this collapses when we first understand quantum theory. Someone who truly understands quantum theory also understands that we have a difficult problem in understanding "reality as it is", and the scientific community which is attempting to explain it to us has been in a serious crisis for over 100 years. By this statement I'm not referring to the practical aspect of quantum theory; it is an incredibly useful theory and is one of the main reasons for the stunning technological development we've seen in the 20th century. However, when we try and understand what it truly says about the world itself, we find ourselves in a state of shock and surprise.

Quantum theory is, as mentioned above, one of the most practical and well-founded physical theories we have. Most of the digital world that developed in the 20th century is in essence a direct result of the accurate predictions of this theory. Personal computers, mobile phones, medical imaging equipment, lasers and almost every digital device known today would not have existed if not for the knowledge of quantum mechanics. The problem isn't the practicality of quantum theory, but rather its implications, and this is what bothers anyone who tries to understand it - how can it be that a theory that allows us to do amazing things in the digital world on a daily basis is so strange, counter-intuitive and simply inconceivable on the most basic level?

Quantum theory began in the start of the 20th century after Max Planck coined the term "energy quantum" as part of the solution for the black body radiation problem (as you may recall, one of the two "clouds" which Lord Kelvin claimed we were left with at the end of 19th century, the solution of which turned human knowledge on its head).

Before getting into the details, it is important to understand that quantum theory explains the behavior of the building blocks of reality. Just as relativity theory explains everything we see in the macroscopic world around us efficiently and consistently, from the gravity of objects on Earth to the behavior of galaxies and of the universe itself, so does quantum theory give as an efficient and comprehensive explanation of all that is microscopic, of the fundamental components of the universe: the particles, fields and waves from which we - as well as everything around us - are made. The more we delve in and understand the behavior of these elementary particles we encounter phenomena, implications and results of experiments stemming from theory which do not match the basic patters with which we perceive the world. In this chapter we will try and describe some of the phenomena unique to the quantum world and their implications for understanding the world. I do not intend to describe the entire theory but rather to focus on the aspects of it which are most relevant to the central subject of this book - time. Eventually, the problems we will discuss in this chapter and the next will help us understand the concept of time, and later on to construct possible solutions to the big questions we've been discussing up until now.

Most of the books I know of which discuss quantum theory usually begin by describing the big question: are the fundamental components of the universe waves or particles? I think that the use of these terms is a cause for confusion and one of the reasons we have a hard time understanding quantum theory. The associations we have with these terms are different from the way they're described in quantum theory. The elementary particles which make up the universe aren't exactly particles, in the sense that they're not what we imagine as tiny balls moving all over the place, and they're not classical waves (as we imagine waves of water or sound) either. As we know today, the fundamental components of the universe are essentially different from anything we know in our macroscopic world, although in many ways they're more similar to a wave, as we perceive it intuitively, than a particle. In my description I will make use of the term "particle" since it is close enough to the truth and can give us a valuable visual image we can understand relatively easily, and also because apparently our language doesn't have terms which can give a better visual idea of what truly happens on a microscopic level. However, I suggest we don't get too caught up in the analogy of particles as billiard balls moving and bumping into one another, since it can be misleading and increase the gap between our perception and reality.

In order to understand what these particles are we have to make an effort and try and imagine an entity which is a combination of an object which is concentrated in a particular point (such as a ball or a speck of dust) with properties of a wave that spreads over a larger space. These two descriptions are contradictory but we will have to make do with them for the moment. Later on, I will attempt to provide a clearer analogy, however right now it's best that we focus on the particle description.

How is the behavior of a particle different from the behavior of a wave? Imagine that you're standing in a pool full of water. You throw a ball which hits the side of the pool facing you. The ball will move in a direct path towards the side and hit it at a certain point. At each and every moment we can identify the exact location of the ball in its path, up to the point of impact. This is classical behavior of a particle. However, if you move your hand in the water and create waves, the waves will move towards the side of the pool and hit each other on the way. When two crests of the waves (the high point of the wave) meet, they will reinforce each other and become a bigger wave. When two troughs of the waves (the lowest point in the wave) meet they will make the unified trough deeper. When a trough meets a crest, they will cancel each other out and no wave will be seen. Eventually, when the waves hit the side of the pool we will see the image familiar to anyone who's created waves in a pool: areas where the wave hits the side at a great height (two crests which have merged), areas where the water hits the side at a lower height (two troughs which have met) and areas where we won't see a wave hitting the side at all (a crest and a trough which have cancelled each other out). This phenomenon of crests and troughs coming together to create a unified shape is called "interference", and is unique to waves - that is, interference is conclusive evidence for the presence of waves.

Why is it important to us to understand if nature's elementary particles are waves or particles? The truth is that this distinction isn't that essential for the understanding of the concept of time. We will discuss it since light and its speed are an essential factor in our discussion of time, and for many years various arguments have been made regarding the particle-like and wave-like nature of light. Furthermore, when we attempt to determine the way in which a particle behaves through experiment, we run into a phenomenon which will lead us to fascinating conclusions regarding the universe and time.

Is light made up of particles or does it behave like a wave? The history of the last few centuries is full of experiments and theories which "proved" that light is a wave and that light is a particle. Each time a theory was presented, supported by a series of experiments, which proved the opposite. During the 19th century, the claim that light is a disturbance in a medium which behaves like a wave became widely accepted, following Maxwell's work on electromagnetic radiation. But then Einstein arrived on the scene and messed up everyone's plans of retirement. In the aforementioned article for which he received the Nobel prize, Einstein proved that light exhibits clear particle-like behavior, and that it is made up of discrete units of energy (quanta of light), which in the 1920s came to be known as "photons". However, Einstein didn't resolve the question of whether light behaves like a wave or a particle, he just made the debate all the more complex.

In order to understand why this debate hasn't really ended to this day we will describe a famous experiment called the "double-slit experiment", whose goal was to examine whether a photon or electron or any other elementary particle is a particle or a wave. The experiment had already been conducted in simpler forms as early as the 19th century and served as proof of light's wave-like nature, but Einstein's discovery (the photoelectric effect) required an improved and up to date version. This is where the true mystery of nature begins to reveal itself.

We take a screen which is sensitive to impact from particles - when electrons or photons hit it, a mark is left at the point of impact. In front of it we place another screen, with a slit in it - which an electron or photon or any other type of particle can pass through (it is the only point on the screen through which the particle can pass). Now we shoot a particle towards the screen, and the mark made on it is the particle's point of impact. That is, much like the ball in the analogy of the pool, the particle moves at a certain path through the slit towards the backmost screen and collides with it at a certain point. This is classical behavior of a particle.

Now we create a second slit in the screen for particles to pass through. If we now fire particles without knowing through which slit each one of them goes through, a pattern of interference will appear on the screen. The particles passed like waves through the two slits (much like waves breaking against a breakwater) and interfered with one another, and this is the image we find on the impact screen. This is classical behavior of a wave.

If so, the particle behaves at times like a particle and at other times like a wave.

What happens if we close one slit?

If we now fire particles which will go through the slit which remained open the interference will disappear and when examining the impact screen, we will only see a collection of points, which shows that we're dealing with particles rather than waves. However, the only difference between the two experiments is our knowledge regarding the particle. In other words, when we know where the particle has passed through, particle-like impact pattern will appear on the screen, and when we don't know where the particle passed through, an interference pattern will appear, hinting at the presence of waves. How can a certain entity (photon, electron, etc.) be both a particle and a wave? A possible explanation is that the particle goes through both slits at the same time, and thus interferes with itself. However, when there is just one slit open and the particle passes through it, it doesn't interfere with itself. But we will soon see that this answer is unsatisfactory.

The problem becomes more complex when we build an experiment in which we know through which slit a particle has passed through even when both slits are open. At each slit we place a detector which beeps each time a particle passes through it, and thus we can know exactly through which slit each particle has passed. When we "fire" particles towards the detector, the interference pattern doesn't appear, unlike the previous experiment, and the screen shows two areas of impact with no interference. That is, **the fact that we know** the location of each particle has changed the behavior of those particles - from typical wave-like behavior expressed through the creation of interference, to typical particle-like behavior. Does our knowledge affect the particles, or perhaps the detectors we place at each slit are responsible for this result? The particles are supposed to pass through each detector, and it's possible that the physical interaction with the detector is what causes the change.

In order to test this, we will carry out an experiment in which our knowledge of the particle's location is indirect. First, we open the two slits and allow the particles to pass through them and create an interference pattern on the screen. Now we place a detector next to just one of the slits which will beep every time a particle goes through it. If the particle goes through a slit with a detector, the detector will beep and we'll know the position of the particle; if the detector doesn't beep this means that the particle went through the other slit and thus we know its position without interacting with it physically. That is, we know that a particle moved along a certain path not because the detector detected it along this path, but because it didn't pass through the detector which attempted to detect it on an alternative path. In experiments of this kind, something amazing is discovered: in these cases as well, the particle behaves like a particle rather than a wave, the same way it did when we measured its position directly. If so, how does the very fact of us knowing its position, without any kind of physical interaction with it, cause it to change its nature from that of a wave to that of a particle?

Another version of the experiment raises questions about time as well. John Wheeler proposed the "delayed choice" experiment. Without getting into the technical details of the experiment, in this experiment it is only after the particle goes through the slits that we decide whether to measure it in a way which should show its particle or wave-like properties; that is, the particle should "decide" the way in which it passed through the slits (as a wave or a particle) only after it happened. And as it turns out, the result is once again surprising: it seems that the particle really does go through the slits in accordance with our decision regarding measurement, which takes place after it's already past the slits. That is, the particle, supposedly, "retroactively changes" its behavior at the moment it passes through the slits, in accordance with our measurement method.

As you've probably realized, the behavior of this particle-wave is different from anything we've seen so far in classical physics, and so it has become one of the biggest riddles in quantum physics and in physics in general.

The main reason, in my opinion, for us being so amazed by the implications of these results, is that our patterns of thought belong to the world of classical physics. According to classical physics the world is made up of basic entities with a clear and distinct existence, and they can be assigned to only two categories: wave or particle. Each entity moves absolutely in space and interacts with entities which are similar to it. It seems that the world does not match this basic pattern of our consciousness, which developed through evolution and which is the source of the language humanity has constructed for itself. Our conscious-linguistic structure, which helped us survive for the past few million years on Earth, doesn't know how to deal with the new microscopic world that's been discovered, which is entirely different from anything we know.

The question of the wave-particle duality of the basic building blocks of the universe is a much deeper question than what is described in the framework of these experiments, and is part of the great debate between two of the greatest physicists of the early 20th century, and pioneers of quantum theory - Albert Einstein and Niels Bohr. Einstein was opposed to the implications of quantum theory to his dying day and believed in Kant's "thing in itself", that is, in the existence of an objective reality. But unlike Kant he sided with the idea that every scientific theory should take the "elements of reality" into account, that is, those entities which make up this objective reality. This is why he couldn't accept a vague entity which is sometimes a particle and sometimes a wave; as far as he was concerned, in order for a theory to be complete, it must describe all of nature's basic entities as they are. This is an ontological view which attempts to explain existing entities in a clear and complete way. For Einstein, the phenomena we described and the ones we will describe later on show that quantum theory is incomplete, and that we need to find additional hidden variables in order to complete it and fully describe the wave-particle entity, doing away with any contradictions.

Niels Bohr, his bitter "rival" in many physical debates, believed in an epistemological approach according to which, following Kant's lead, we cannot talk about "reality in itself", but only about what our consciousness perceives. For this reason, Bohr promoted an approach called "complementarity", which allows us to accept contradictory insights regarding the world, as complementary understandings of reality. According to Bohr, the fact that we perceive basic elements in the world as containing a wave-particle duality or as contradicting one another is an insight which should be accepted as the complete image of reality. In developing this approach Bohr was influenced by eastern philosophy, which views reality as a unity of contradicting viewpoints, as part of the one complete reality - a view expressed in the yin-yang symbol of opposites which complete each other, as we've already mentioned.

Bohr's view, which accepts the dual and contradictory nature of the foundations of quantum theory, has won over many supports over the years, and is now a matter of general consensus among the scientific community. The acceptance of this approach - which is called "The Copenhagen Interpretation" after the place of residence of Bohr and his team - doesn't stem from delving into its philosophical implications, which leads to a very vague understanding of reality, but from a positivist stance. In order to keep working in a lab or a particle accelerator, you don't really have to get caught up in the philosophical implications of quantum theory, and Bohr's stance allows the physicist to enjoy its practical aspects and ignore the contradictions it involves.

The debate between Einstein and Bohr is a fascinating story of thought experiments and an epic clash of minds which went on for years, and it will accompany us throughout the story of quantum theory.

"Dead" and "Alive" - Schrödinger's Cat and Equation

When we describe the behavior of a particle, we use the "wave equation" developed by Erwin Schrödinger in 1926, with which we can know where a particle is at any point in time. And yet, whenever we're not making a measurement of a particle and the particle doesn't have a definite and known position, what we get from the equation is the probabilities to find the particle in various locations when we do perform the act of measuring. The equation gives us a kind of wave of probabilities which describes the points in space in which we might find the particle, and the probability of finding it in each of these points. In other words, we can only know the probability of finding a particle at any given position and we can never predict for certain where it will be. Our inability to know the position of each particle in space stems from the fact that the particle has no definite position, and not because we're lacking information (later on we will discuss various interpretations of quantum theory known as "hidden variables theories", which claim that the particle does have a definite position but we don't know what it is). The meaning of this is that the building blocks of universe have no definite position before we measure it. We can imagine the potential of each particle's position as a "cloud of possibilities": it is very grey in areas where the probability of finding the particle are highest, it is white where the probability is lower and invisible where the probability is almost non-existent.

When we perform a measurement, we find the particle in a definite position within the cloud of possibilities. At the moment of measurement, we find a single, specific point, out of all the possible points with varying probabilities which appear in the wave equation, in which we find the particle. This "cloud of possibilities" collapses to a specific point and the particle receives a clear physical location. This is the famous concept of the "collapse" of the quantum wave equation: the wave equation which describes the various probabilities of the particle's position "collapses" into a single result. The "collapse" is random and so there's no way of accurately predicting the particle's location before we measure it. The probabilities and the component of randomness which have entered quantum theory led Einstein to declare in a letter he sent his physicist friend Max Born in 1926: "Quantum mechanics is certainly impressive, but an inner voice is telling me that it's still not the real thing. The theory says a lot but does nothing to bring us closer to the old man's [God] secret. I, in any case, remain convinced that he does not play dice".

Einstein, who as we know was a determinist, could not accept a random component in nature. With his famous statement regarding God and dice he's essentially saying that each entity in nature has a clear and distinct location, and it's simply not possible that God included a random and probabilistic component to the physical properties of objects in nature. Bohr's response to Einstein's words was: "Don't tell God what to do with his dice".

The big question is where the particle is before we measured it. What does it mean for a particle to be in a "cloud of possibilities"? It either has a position in space or it doesn't; a particle can't be in several locations at the same time. The surprising answer is that the particle is located at all the possible locations allowed for by the cloud of probabilities. Such a location, which actually includes several simultaneous locations, is called "superposition".

Going back to the "double slit" experiment, what goes through the slits is actually the "cloud of possibilities" made up of the wave equation's probabilities. When we allow this "wave of probabilities" to go through just a single slit, or when we place a detector in its path, we can know the particle's position inside this "cloud" and thus we're performing a measurement on the particle. Since a measurement is taking place the equation collapses and we find the particle in a certain position, and thus it hits the screen as a particle, not a wave. If we don't know through which slit the particle passed, since we didn't perform a measurement, it goes on behaving like a "wave of probabilities" and so appears on the screen as an interference pattern.

Even though the process is clearer now, we still remain with the big questions: how can it be that the same entity behaves like a particle at times and like a wave at other times? And how does the act of measuring change its wave-particle nature? Before we keep discussing the question of measurement, I have to emphasize another thing which is mostly missing in these descriptions in popular science literature since it complicates things even more. When we imagine interference, we see in our mind particles going through the slits the way waves at sea go through a breakwater, hitting each other and interfering similarly to the waves in the pool I described earlier. However, in our case the analogy is even more surprising. When we send a particle through two open slits, that is, in a state of a wave, we see an impact point on the screen. Every additional particle we send will create an additional point on the screen. The interference pattern appears on the screen when many points representing particles accumulate. What appears on the screen is a collection of points arranged in a way that matches the probabilities of each position of the particle, in separate lines of interference. That is, the interference pattern is that of a wave of probabilities and doesn't express two interference between the particles themselves - since if the particles appear as separate points and do not interfere with one another, then what's actually doing the interfering is the wave of possibilities; the interference pattern will appear only after many particles in different locations appear, in accordance with the wave of probabilities.

When the particle has particle-like properties - for example when a measurement of its position is carried out before it reached the screen - we will see an accumulation of particles on the screen in a round concentration at a certain area, with no interference pattern. The amazing thing is that what's going through interference is the probabilities, and not the particles themselves. Anyone who's still conscious and managed to follow all that, understands that the world described here is simply inconceivable in terms of intuition, and is unlike anything we know from our everyday lives in the world.

Erwin Schrödinger, who had quantum theory's fundamental wave equation named after him, was an Austrian physicist, Nobel laureate and contemporary of Einstein and Bohr. Schrödinger was an open minded man, who opposed the Nazism in Germany and so left it in the early thirties. If you visit the moon as tourists one day, you'll find out that after his death, a large crater at the far side of the moon was named after him.

Schrödinger, as a friend of Einstein, didn't like the probabilistic aspect of quantum theory. A famous thought experiment he came up with, known as "Schrödinger's cat", was meant to show the paradoxical results which follow when you attempt to apply the "Copenhagen interpretation" to the transition from the microscopic world to the macroscopic one. What happens when we try and understand how the duality and probabilistic nature of the particle world affect the world of objects which these particles make up? Schrödinger describes a cat who's inside a closed box, with a device containing a single atom of a radioactive substance. As soon as the atom decays, a mechanism is activated which will release a gas that kills the cat. The atom's decay takes place in accordance with the predictions of quantum theory and can take place with a probability of 50% in a certain period of time. After this period of time has passed, so long as no measurement was carried out, according to the wave equation the atom is still in a superposition or a "cloud of probabilities", where there is a 50% probability of the cat still being alive and a 50% probability of the cat being dead. In accordance to what we knew about the double-slit experiment, the wave equation and the Copenhagen interpretation, before we carried out a measurement the cat is still described by the wave equation, that is, the cat is still in a state of superposition: 50% alive and 50% dead. Since the equation hasn't collapsed into one of the two possible states (since we haven't carried out the measurement yet), the cat is simultaneously both alive and dead. This is the consequence of applying the insights of the Copenhagen interpretation to the world of people, cats, and objects made up of many particles. When we open the box the wave equation will collapse into one clear and distinct result and we'll be able to see if the cat is alive or dead; however according to to quantum theory, before we carry out the measurement the cat is **physically** in a state of superposition of "alive" and "dead". This conclusion sounds absurd of course, and this was exactly Schrödinger's goal - to demonstrate the absurdity of this description. But this is the significance of the statement that a particle doesn't have a definite, well-defined property before we've measured it.

And this is where the problem lies: this strange story is the story of the building blocks which make up each and every one of us.

What do their indefinite properties say about us and our world? Our journey into the deep mysteries of the universe continues.

Measurement, Consciousness, and Parallel Universes - The Interpretations of Quantum Theory

We've mentioned the act of measuring and its influence on reality according to quantum theory quite a lot now, but we haven't tried to understand what a measurement actually is. A measurement is a critical action which constitutes the turning point in the process the particle goes through and in our understanding of its behavior. At a first glance, it is clear to us what a measurement is when we place a detector in the particle's path or when we open the box and look at the cat. But if we attempt to delve into the concept of measurement, we find one of the greatest riddles of the process a particle goes through. The act of measuring, according to the accepted interpretation, is what causes the collapse of the wave equation to a single result from a collection of theoretical probabilities, and the transition of the system being measured from one state to another. As I've shown above, in certain situations it can be claimed that the measurement creates a physical interaction with the particle and causes its collapse. For example, when we place a detector in the particle's path, we can claim that the physical interaction between the particle and the detector is what causes its collapse. However we've already seen that the equation collapses even when a measurement takes place that doesn't involve a physical interaction between the detector and the particle - this is what follows from the experiment in which we placed a detector in only one of the slits and indirectly concluded the location of the particle from the fact that it **didn't** pass through the detector.

At what point does the measurement take place, determining the location of the particle? Using the example of Schrödinger's cat: when does a measurement take place, resulting in the cat going from a state of "dead"/"alive" superposition to a single defined state? Prima facie it would seem that the equation collapses when we look inside the box. Is a human observation enough to cause the collapse of the wave function to a specific position or does the collapse take place at an earlier stage? Does it happen when the measuring device performs its operation? Is it when the result appears on the measuring device? Is it when the information arrives at the retina of the human eye or when our consciousness comes into play? The act of measuring is a chain of actions and indications, and in order to understand physically when the act of measuring takes place, we must point at the precise physical state which causes it. Where do we cut this chain of events and decide that that's where the measurement takes place, and the wave equation collapses?

Phrasing the question in such a way confronts us with a difficult problem, since on one hand we're trying to find the objective physical explanation for the collapse, but on the other we're bringing the observer into the measurement process. The observer is part of the process, but he must be external to it in order to preserve the possibility of understanding the process of collapse objectively. As soon as we've included the observer and their complex consciousness in the process we've created a much more complicated chain of events, part of which isn't fully understood by science (consciousness) and which has subjective components. The tension between the physical process and human intervention in the processes is what led to the flourishing of the view, common in certain spiritual doctrines nowadays, that views quantum theory as a decisive proof for the effect of human consciousness on processes in the universe. And indeed, over the years various theories were raised, such as the one by Eugene Wigner, which emphasize the significant role of consciousness in the process of collapse.

In my view, all these questions are moot at this point, since measurement is an infinite process of events which begin with the measuring device itself, go on to involve our senses and the eye's neural system, which receives the information from the measuring device in the form of photons, and end with complex processes in the brain and consciousness, and right now we don't have any way of proving when in the measuring process the wave equation collapses.

There's also the question of whether the human factor and its consciousness are a necessary condition in the particle's process of collapse. If so, what happened in the universe before the first human was created? (In the spirit of "if a tree falls in the forest and no one's there to hear it, does it make a sound?") or as Einstein put it: "Do you truly believe that when you're not looking at the moon, it doesn't exist?"

Beyond the measurement problem and as part of a deeper understanding of quantum theory, Werner Heisenberg formulated the "uncertainty relations", which define a fundamental uncertainty in nature. According to this principle we don't have the ability to know certain properties of particles at the same time. For example, we cannot know the exact location of a particle and its momentum (velocity and direction of movement) at the same time. The more precisely we know one parameter, our knowledge of the other one becomes less precise. According to the popular interpretation this uncertainty is objective in nature, that is, the particle truly doesn't have both an exact location and momentum, and our inability to know both these parameters at the same time isn't a result of some technical limitation or lack of knowledge on our part.

The description I've given here regarding wave-particle duality, the measurement problem and the uncertainty relations is based on the "Copenhagen interpretation", which is the most widely accepted interpretation of quantum theory today, as I've mentioned. However, in the last hundred years several alternative interpretations were suggested in an attempt to find different physical explanations for the phenomena we've described. I will describe the three most important sets of such interpretations.

The first set is called "hidden variables theories": these theories attempted to add components and variables to quantum theory which will complete the information we're supposedly lacking in order to make the duality, internal contradictions and the uncertainty go away. The "hidden variables" interpretations were created as a result of the desire to preserve the classical causality which is so deeply embedded in our minds and perception of the world. All the scientific theories which we've known about up until quantum theory, including the theory of relativity itself, are classical causal theories. These theories present a world in which a complete knowledge of a given physical state allows for a certain computation of any later physical state; everything has a cause and there is no essential limitation stopping us from knowing what that cause is. As soon as we have all the information about a certain physical state, nothing will stop us from predicting what will happen to a system we then choose to examine. It is possible that for technical reasons not all the data will be available to us, however theoretically, if not for the technical limitations we could've had a complete knowledge of the world and calculate what will happen next. Quantum theory has shuffled these deterministic cards and included random and probabilistic components into nature itself. Now we have in our hands, for the first time in the history of science, a theory which doesn't allow us to know a physical state at any given moment - not because of technical limitations but because the world isn't deterministic but is rather probabilistic and random. For this reason, the theory doesn't allow us to predict exactly what will happen next. This insight has huge implications regarding our understanding of nature and our place as human beings with free will living in the world. When we add randomness to nature, we lose the ability to make decisions and control outcomes in the world. Classical physicists couldn't accept the death of classical causality and attempted to find variables which remained hidden in quantum theory in order to complete our information regarding the physical state and allows us to remove randomness from the scientific description. And yet, the hidden variables theories were not free of faults themselves, and none of them could serve as a worthy adversary for the "Copenhagen interpretation".

A second type of interpretation doesn't try to add variables to quantum theory, but instead tries to solve the measurement problem by removing the collapse of the wave equation from physical descriptions. According to this theory, no collapse takes place at the measurement stage and the different possibilities which lie in the "cloud of probabilities" take place at the same time in parallel worlds. The "parallel worlds" theory was developed from an idea raised by Hugh Everett back in 1957, and which received reinforcement from the scientific community over the years. According to this theory, as soon as a measurement take place, the wave equation doesn't collapse into a single result but rather goes on to describe the physical state, since each possibility that existed in the wave equation's "cloud of probabilities" goes on existing in universes which are separate and parallel. At every point where we'd once identify a "collapse", there is a split of our world into several separate universes, in which each of the various possibilities is realized. At the same time, our consciousness also splits in each universe to a separate consciousness which is only aware of the possibility which was actualized in that universe, and so we never experience more than one possibility in our reality. If we take "Schrödinger's cat" as an example, then according to the "parallel worlds" theory, as soon as the box is opened the universe splits into two: in one universe the cat is alive, and this is what we see and are aware of when we open the box; in a parallel universe which has split off from us, the cat is dead and that's all we know. Everett and his followers claim that we're actually in a superposition of all of these possibilities even **after** the act of measuring, it's just that now the superposition is a combination of all the worlds existing in tandem. The cat goes on being both alive and dead in parallel universes.

In principle, this suggestion solves the measurement problem since it spares us having to deal with the murky and random concept of the "collapse of the wave equation", and it does away with the random element which lies in the realization of just one possibility out of a set of different probabilities. However, this view raises three serious problems:

1. The parallel worlds theory doesn't offer a real solution to the measurement problem since eventually the split takes place as a result of some measurement, and even if we accept the theory, we still need to explain what causes this split. Furthermore, the theory doesn't solve additional problems such as the question of duality and uncertainty.
2. The theory doesn't explain why the realization of the various possibilities matches certain probabilities. For example, if there were three possibilities according to to the wave equation, each with a different probability of coming true, what is the significance of these probabilities if each one of them is realized with a probability of 100% in another universe?
3. I've already presented my criticism of parallel worlds when discussing the arrow of time. As you might recall, a similar explanation was already offered as a response to the question why our universe is in a state of low entropy. This explanation, which is called the "multiverse", is very similar to the "parallel worlds" of quantum theory. Many physicists create a link between these suggestions, which have been made separately but which can provide an answer to many

> questions on the quantum and cosmological level, and which could form a part of a unified theory. In my opinion this solution is too easy, and its main problem is that one can't help but wonder how scientific these proposals are. As I've described in the discussion on the arrow of time, we can't disprove these theories as long as we don't think of a way to examine the theoretical existence of such parallel worlds at the very least, and so they cannot be considered to be scientific theories according to Popper's principle of falsification. In addition, according to to the famous philosophical argument known as "Occam's razor", we cannot accept such a "wasteful" theory, which requires an almost infinite number of parallel universes in order to provide an answer for a phenomenon we see in nature. The explanation should be much simpler, and nowhere near as wasteful.

A third interpretation for quantum theory attempts to explain the gap we experience between the phenomena discovered in the microscopic world and the world of large objects, or in other words, to explain why we do not see cats that are both alive and dead at the same time in our daily lives. Solutions to this question have been raised for years, but it was only in the 1980s that Ghirardi, Rmini and Weber (or in short, GRW) offered an organized theory called "Spontaneous Collapse Theory". This theory doesn't try to dodge the concept of the wave equation's "collapse", but rather claims that the collapse occurs on its own, spontaneously. Once every few million years, a spontaneous collapse of a particle to one result out of a "cloud of probabilities" it is situated in can take place. This spontaneous event is very rare and so when we're looking at just one particle the probability of seeing such an event is infinitesimally small. However, when we're looking at a macroscopic object such as a book or a table, which is made up of an astronomical number of such particles, the probability of a spontaneous collapse taking place in at least one of these particles is 100%. The collapse of one particle, which is interacting with the rest of the particles which make up the macroscopic object, will immediately cause the collapse of all of them, and thus the collapse of the object as a whole, and it will go out of a state of superposition into a state of a singular result and singular position in space.

GRW's proposal is very strong and provides an excellent explanation for the fact that we do not see quantum phenomena in everyday life, where we only encounter objects which are made up of a huge number of particles. This view is also compatible with our current understanding that the quantum world is characterized by de-coherece: an individual particle is not truly separate from its environment and so is always in a state of interaction with other particles. Even a particle moving through space collides, every now and then, with a photon or some other kind of particle which is moving through space. Every such collision is a kind of measurement which causes the particle to transition from a state of superposition to a specific location. Since all the objects in the universe are constantly being "bombarded" by various types of particles moving around the universe, measurements are taking place at every single moment. That is, when we're looking at the world we cannot ignore the fact that it is made up of very complex objects, and that the particles that make up these objects are in a state of infinite and incessant interaction with each other and with particles which are constantly colliding with them. In such a state we're no longer talking about an isolated particle in laboratory conditions, and so the phenomena in the "real world" do not match the quantum description. It's important that you remember the concept of spontaneous collapse and de-coherence; we will get back to them later on after we've become acquainted with complexity theory and attempt to find a way to connect it with quantum theory, two theories which may be different but which I think have very interesting points of contact.

One of the questions which spontaneous collapse theory and de-coherence don't solve is why, as part of the process of collapse, one of the possibilities out of the various probabilities is chosen, rather than another. The component of randomness when it comes to the path the system chooses remains unexplained. Further on I will propose a model based on network and complexity theory which attempts to link between this phenomenon and another quantum phenomenon, which we will discuss in the next chapter (quantum entanglement), in order to provide an answer for the question of randomness, or more precisely, to the question of why a certain possibility out of an entire "cloud of possibilities" is realized in reality, rather than another.

I should also point out that the phenomenon of de-coherence should bother anyone who's interested in applications of quantum phenomena in the real world - whether it's advanced scientific applications such as quantum computers and use of quantum phenomena to create and decrypt codes, or popular applications of quantum theory which attempt to link between quantum collapse and our ability to make personal choices in the world (in other words, anyone who's seen movies such as "The Secret" or "What the Bleep Do We Know?!"). De-coherence is one of the biggest obstacles when it comes to applying quantum phenomena in everyday life, since it raises a doubt regarding our ability to use the superposition of particles in order to make decisions or carry out computations without an immediate, uncontrollable collapse taking place.

As we now realize, the big questions which follow from quantum theory remain unsolved, and still make fertile ground for debate in both the physical and philosophical worlds. In my opinion, the language we use is a critical element both in understanding the questions we're dealing with and in the attempt to provide a visual explanation for physical reality. I'm convinced that our language is lacking the tools for a visual description of what goes on in the sub-atomic world, and that the use of terms such as "particle" or "wave" is misleading and impairs our ability to construct a satisfactory explanation when it comes to our intuition. Quantum theory and its implications seem illogical to us only because the language we use to describe them is a human, limited language, and since reality itself owes nothing to human intuition.

The Collapse of the Arrow of Time - Measurement, Collapse and Backward Causation

The attempt to understand the world through quantum theory raises a good amount of bewilderment. The philosophical and physical implications which follow from the theory have been sparking intellectual debates and discussions on the nature of the universe we're living in for many years now. Our difficulty in fully understanding all the implications of quantum theory is a source of much confusion, but also paves the way for different and perhaps strange ways of thinking, regarding the world as a whole and time in particular.

Quantum theory does not directly deal with the concept of time, unlike relativity theory. However, quantum theory includes various sparks of insight which can be used in our discussion on the concept of time, particularly to better understand time, and maybe even pave the way for theories and conclusions which directly involve the nature of physical time in the universe.

As mentioned above, Schrödinger's wave equation is the basis for the description of a particle's movement in space and time. As long as we haven't performed a measurement and the particle hasn't "collapsed" into a certain location, the equation describes the movement of the entire "cloud of possibilities" in space and time. In this sense, the wave equation is a classical equation of motion and is no different from any other equation of motion which describes the movement of objects in the world, except for the fact that it describes different possibilities moving in tandem, rather than a precise, specific location of a particle moving in space. But this isn't where the story of the wave equation ends. At some point a "collapse" will occur, either spontaneously or as a result of a measurement, or a split between universes will take place. Whichever interpretation you choose, there will be a point in time in which the "cloud of possibilities" will disappear and we're left with a specific location for the particle which is moving through space and time just like a classical object. As we've seen in the discussion on the arrow of time, time is inherently symmetrical and the equations of physics don't differentiate between past and future. We see a distinct arrow of time because entropy in the world is always growing, solely due to probabilistic reasons, and not because the arrow of time is some fundamental law of the universe. The act of collapse is unique, since the collapse of the wave function introduces an irreversible arrow of time into the microscopic world which cannot be explained with Boltzmann's statistical entropy. For this reason, it seems that the collapse of the wave function is more fundamental when it comes to the laws of the universe than statistical mechanics, which explains the thermodynamic arrow of time. Since the collapse is irreversible, immediately after it the information included in the wave equation regarding the different probabilities is gone and cannot be recreated.

Since the measurement problem and the collapse of the wave equation in quantum theory are a strong expression of the arrow of time, and since the information which is lost from the "cloud of possibilities" cannot be recreated, it's implied that when we measure a quantum system, something happens which takes the system into an irreversible state. For this reason, quantum theory is revealed to be a theory which is not symmetrical in time, unlike other physical theories, which are clearly symmetrical. Quantum theory is an expression of a distinct arrow of time pointing in just one direction (the future), the same way we experience the world intuitively, and it's possible that this hints at the fact that quantum theory holds the key to solving the riddle of time.

Another term we discussed, and which raises interesting questions regarding the concept of time, is the "delayed choice experiment". As you may recall, in John Wheeler's experiment, the choice on how to measure the particle in retrospect influenced the way the particle passed through the slits "backwards in time". The choice of the person conducting the experiment has a supposedly future effect on the particle's state in the past. This conclusion is of course a very controversial one, but it serves as yet another reinforcement to the widely accepted claim, whose main representative is an esteemed philosopher of physics called Hugh Price, regarding the existence of "backward causation" in the universe. Using very complex but also very convincing arguments, Price claims that a large part of the embarrassment around quantum mechanics today stems from deep-seated and unfounded intuitions regarding the temporal asymmetry of the microscopic world, which have been created as a result of our understanding of the concept of "collapse".

Classical causation is the causal link between cause and effect we see in our everyday lives: every action in the past causes a certain effect in the future, and this is always the way of things. We don't experience the influence of causes on effects **in the past**, we experience causation as a process which begins in the past and which affects the future. This, of course, is a law which is asymmetric in time. Price claims that the universe is symmetrical when it comes to causation as well, that is, there is such a thing as "backwards causation", which is the influence of a certain action in the present on the past and not just the future. Backward causation is not reverse causation as is seen in the video clip played backwards. In backward causation, when we "play back" an experiment where a particle is measured, it will return from a state of being in a specific location on the screen to a state of a "cloud of possibilities" before the measurement has taken place. Price claims that this isn't what happens in practice: when the factor influencing the system comes into play, it affects both the future state of the system and the past state of the system, before the measurement took place. In other words, when I perform a measurement on a particle, I influence both its future state (collapsing into a specific location) and its past state. From the moment of measurement, the particle's past also collapses into a specific location, and so the act of collapsing is symmetrical in time. In the delayed choice experiment, performing the measurement causally affects the state of the particle in the past, at a point before it went through the slits. That is, the future can affect the past and there is full symmetry when it comes to time.

At this point Price follows Kant and claims that the causality we see in our everyday life is the result of a pattern in our consciousness rather than a law of nature existing in reality itself. We do not see the symmetry in time due to the structure of consciousness and the asymmetrical pattern through which we see the world. In the microscopic world, we see an action which we can't explain due to the limitations of our consciousness, but it is hinting at the symmetry of time. The supposed "collapse" of the wave equation is not an action in the physical world but rather a process which takes place in our consciousness alone, stemming from the asymmetrical lens through which we see the world. The "collapse" takes place in our minds only and hides the fact that in the real world there is no difference between past and future when it comes to causation.

Backward causation may be able to explain the process of collapse of the wave function and some of the bizarre phenomena we see in quantum theory, but it is of course a philosophical, and very unintuitive view, and I'm not sure if it can be disproved in any way. In any case, this is an insight which brings back the symmetry which took a hit due to the concept of collapse, and symmetry is one of the properties which most physicists would say are the most fundamental to the universe. It is important to mention here one of the most esteemed Israeli physicists, Prof. Yakir Aharonov, who is seriously and physically examining the quantum effect of the future on the past. Aharonov developed measurement methods called "weak measurements", which indicate that in the quantum world, future states influence the events which take place in the present and even the past.

If so, quantum theory presents a world moving along a classical arrow of time with irregularities stemming mostly from the act of measuring and the collapse of the wave equation. Furthermore, it includes hints regarding the possibility of a bi-directional arrow of time, and backward causation, as in the delayed choice experiment. However, quantum theory does not include a direct and clear statement regarding the nature of time since it does not discuss it directly.

The understanding of quantum theory is important to our discussion on the concept of time in this book mostly because later on, when we attempt to create a clearer explanation for the big questions, we will be using the language and rules of quantum theory. Furthermore, the foundation of understanding we've built in this chapter leads us to another, even more important quantum term, which stems from quantum theory. In 1935 Albert Einstein, Boris Podolsky, and Nathan Rosen published a claim which was supposed to prove the implausibility of quantum theory. This argument, called the EPR Paradox (after their initials), introduced the term "quantum entanglement" to the world, one of the most astonishing concepts ever created in a physical theory, whose implications completely change, even as this book is being written, our perception of reality.

CHAPTER 8:

The Hidden Universe - Quantum Entanglement

The Spooky Connection - Quantum Entanglement

In this chapter I will describe an attempt to prove the implausibility of quantum theory which introduced the world to one of the least known, and most significant concepts, in my opinion, for understanding physical reality: quantum entanglement.

As I pointed out several times, Albert Einstein may have been one of the heralds of quantum theory, but things didn't turn out exactly the way he'd hoped. To his last day, Einstein believed that the theory is incomplete and that we need to find solutions for the difficult questions that arise from it, in the form that became widely accepted by the physics community at the time. In order to prove the implausibility of quantum theory, Einstein and his colleagues came up with various thought experiments, which describe hypothetical situations which should result from a practical application of the mathematical theory. Such practical experiments were not possible at the time due to technical limitations, they were either too hard or too expensive to carry out. The results of the thought experiments, according to the mathematical theory, were supposed to be absurd and fly against the rules of logic. The discussions which developed following these thought experiments, mostly between Einstein and his "rival" Niels Bohr, accelerated the development of quantum theory during those years.

The crux of the debate between Einstein and Bohr, two people who can be simplistically described as representing opposing views when it comes to interpreting quantum theory, revolved around the probabilistic component of quantum mechanics. As I described in the previous chapter, quantum mechanics (using Schrödinger's "wave equation") only allows us to know the probability that a certain property of a particle (position, momentum, spin, etc.) will receive a value, and this value is only received once the measurement itself is carried out (the "collapse of the wave function"). Thus, for example, until the position of a particle is measured, we cannot know where it is and in essence it has no specific location in space, and so we only know the various probabilities for finding it in different locations (this is the essential difference between quantum physics and classical physics, which states that we can accurately predict where a given object will be at any point in the future, based on its current position and velocity. (In quantum physics we have no such knowledge regarding position, only knowledge about probabilities).

The heart of the dispute between the disciples of Einstein and those of Bohr is the question of the source of our lack of knowledge regarding the particle's properties; does our lack of knowledge stem from us lacking information, i.e., the theory is incomplete (the "hidden variables" approach of Einstein and others), or is it because before the measurement has been carried out, the particle doesn't actually have a defined position and the act of measuring is what causes the "cloud of probabilities" to collapse to a specific location (Bohr's "Copenhagen interpretation").

There are many stories about the debates between the two, who were close friends and bitter professional rivals. One story tells about conferences and conventions where one of them would get up on the stage, propose a thought experiment which contradicts the view of his rival and send him running around like a man possessed back and forth between his physicist friends until he'd find a solution to the riddle; after the solution was presented, the other man would enter a prolonged bout of solitude at the end of which he'd present a claim whose truth he's convinced of, and so on and so forth. These discussions between the two became the stuff of legend, and mostly did a lot to promote the knowledge of the physicist community within a relatively short period of time.

The gap between Einstein and Bohr's world views mostly focuses on the basic definition of the physical world. Einstein advocated a realistic world view, which sees the physical world as a real thing, which exists in itself and ascribes the objects in the world an independent existence and objective properties. According to this world view the real world exists outside of us and it has clear, defined properties regardless of whether we're observing them or not. So when Einstein was constructing his physical theories he searched for those "elements in reality" unrelated to the observer observing them. Bohr, on the other hand, believed in a positivist world view which contradicts realism. According to this view, scientific study of nature should only focus on what we perceive with our senses and should not make conclusions about an objective nature with an existence that our senses don't have access to. That is, according to Bohr, we can't say that the world outside of us exists and has properties which we perceive directly with our senses, and so anything we can say about the world is limited to the framework of our sensory and conscious perception of it. This view places a lot of importance on the observer, since anything we can say about the world depends on what the observer can sense and perceive. If so, when all is said and done, there is no meaning to the existence of phenomena in nature if there's no observer observing them.

One of Einstein's most famous quotes, "God does not play dice", is an expression of the view which claims that the real world simply cannot be based on nothing but probabilities. This sentence also expresses Einstein's vehement opposition to the idea that nature has objects or entities which do not have a defined existence or defined properties before we've observed them, and so it's impossible for a particle, for example, not to have a location before it's been observed.

Einstein's belief that it's impossible for a particle not to have defined properties before it's been observed, led him in 1935 to publish the "EPR Paradox" together with Boris Podolsky and Nathan Rosen which I mentioned at the end of the last chapter.

The EPR Paradox deals with a quantum phenomenon called "entanglement". A state of entanglement is created by an interaction between two or more particles, as a result of which these particles can be viewed as a single, entangled system, or in other words - they can be described with a single wave equation. The technical way in which this entanglement is created is irrelevant to our purposes, however it is important to understand that entanglement takes place in nature on a regular basis, and it can even be created artificially in a lab. The story I'm about to tell you today is complex and very hard to grasp intuitively since it has no parallel in everyday life. In order to simplify the story, I will use an analogy that likens the particles to a pair of twins. The twins have left their mother's womb together and from the moment they were born, regardless of the distance between them, there is a certain unity between them that makes it possible to think of them as a single system. However, these twins go to different schools which are far away from one another; every morning each one of them goes to a school at opposite sides of town and they have no technical way of communicating between them.

Up until now we've talked about how the wave equation describes the particle and its properties as a "cloud of probabilities" until a measurement or "collapse" takes place. Once the measurement is performed, we receive a singular result, in accordance with the probabilities which appeared in the wave equation. As we mentioned, in an entangled system there is a certain connection between the two particles which allows us to describe the two of them using a single wave equation. Each particle has its own "cloud of probabilities" and the probabilities appear in the unified wave equation. Put simply, we now have two particles and a certain way of predicting the probabilities of each particle collapsing to any result that can be received in the event of a measurement. Back to our twins: each twin arrives in school every morning and buys a sandwich at the vending machine. Each machine makes a recommendation every morning, for one sandwich out of three possible kinds: cheese, tuna, or chocolate. The recommendation is random and the probability of the twin accepting the recommendation or choosing another sandwich is equal, that is, there is a 50% chance of the twin accepting the recommendation and a 50% chance of him rejecting it. This is the twins' wave equation (the act of making a recommendation is analogous to performing a measurement in the quantum world).

Quantum theory predicts that in an entangled system, performing a measurement on one of the particles will immediately affect the state of the other particle. When the two particles are described in a unified way and we measure one of them, a strange correlation appears between the measurement of this particle and the collapse of the second one. For example, if we measure the position of particle A, particle B also collapses to a result which corresponds to a certain position. The problem is that the results we get for the two particles do not match ordinary probabilities, and the correlation between the two results is much higher than what is expected according to the probabilities. In the analogy of our twins, the significance of the correlation is that each time the two machines recommend the same sandwich, the two twins will give an identical answer: they either both accept the recommendation, or they both reject it. This correlation does not match regular probability, according to which each twin still has a 50% chance of accepting the recommendation or rejecting it. The correlation between the two twins or the two particles in quantum theory cannot be explained by the theory itself.

To summarize, in a state of entanglement we can describe two particles using a single wave equation, and when one of them collapses due to a measurement of one of the particles, we immediately receive information about the second particle as well, even if we haven't observed it. Going back to the twins: if we know what the reaction of one twin was to a certain recommendation, we can immediately know what the reaction of the second twin will be to the same recommendation.

Einstein, Podolsky and Rosen identified this problem in quantum theory and proposed two possible solutions for it. One solution is that there is a certain correlation **in advance** between the two particles, and this correlation is what quantum theory needs to explain. If a measurement of one particle's position immediately gives us the position of the second particle as well, then the correlation between the particles and the properties of each particle were predetermined. In the case of our twins, this means that the two twins decide together each morning, before they go to school, how they will react to each recommendation. This coordination between the two explains the precise correlation between their reactions. And yet, if measuring the position of one particle immediately gives us information regarding the position of the second particle, then we can measure the momentum of the second particle and then theoretically we will have precise information about both the position and momentum of the second particle. Since in this way we can conclude both the position and momentum of the second particle, these properties are then elements of reality. According to quantum theory there is a fundamental uncertainty in nature and these properties are not elements of reality, the assumption of a predetermined correlation between the properties contradicts Heisenberg's uncertainty principle; and so quantum mechanics contradicts itself and is therefore incomplete. If we can predict the result of a measurement which will take place in the future with certainty, without interacting with the system prior to the measurement, then the result already amounts to a fact; the result is an element of reality as early as the prediction stage. According to the EPR Paradox, if the empirical predictions of quantum theory are true, then the world includes elements of reality which do not match the elements in quantum theory's description.

In other words, since the existing theory cannot explain the correlation between the particles it is incomplete, and we must add variables to it in order to explain this phenomenon. Here we return to the "hidden variables" theories I presented in the previous chapter, whose goal is to fill in the blanks in the incomplete quantum theory.

A second solution proposed by Einstein, Podolsky and Rosen to the problem of correlation between the particles is that there is a connection and transfer of information between the particles which leads to the second particle "knowing" what happened to the first particle and responding accordingly. Back to our twins: the first twin who answers the question somehow conveys his response to the second twin, who immediately responds in full accordance to the response of the first. However, the influence must be immediate, and herein lies the problem. According to what we know so far in physics, two objects can only influence one another through physical interaction or through some transfer of information between them. This transfer of information must be local and take place at a speed that does not exceed the speed of light, which according to relativity theory, is the highest speed in nature.

In quantum entanglement the measured particle affects the second particle immediately, regardless of the distance between the two - despite the fact that if they are light years away from one another, the effect should only take place after a few years. Since according to this solution, the transfer of information is immediate and much faster than the speed of light, according to Einstein, this solution is impossible (he even called it "spooky action at a distance"). And so, we're left with the other possibility: quantum theory is incomplete and must be replaced with a more complete theory.

Out of sight, out of mind - Bell's theorem and Aspect's experiments

Ever since the paradox was published, many have tried to solve it, but since technologically speaking it was impossible to conduct an experiment which would verify EPR's conclusions, the discussion was abandoned to a certain extent in favor of further development of the theory for more practical needs.

Over the years, as I've already mentioned, several "hidden variables" theories were proposed which attempted to explain the strange results of quantum theory using a more complete theory. These theories also attempted to explain the EPR Paradox and provide answers to the question of the unusual correlation between the distant particles. As I mentioned above, the EPR Paradox is caused because of the assumption of locality, that is, the assumption that the effect of one object on another cannot be faster than the speed of light. Hidden variables theories tried to find the variable which would be able to handle the results of the paradox while preserving the principle of locality. Following these theories' lack of success to provide a complete and consistent answer to the big questions, in 1964 John Stewart Bell published an article which proposed a mathematical theory which made it possible to experimentally examine the question of the correlation between the particles and the non-local effect beyond the speed of light. According to the article, if we assume that the world is local (that is, that there is no immediate effect at a distance) we can create a formula which describes a state of quantum entanglement (the "Bell inequality") which is contradicted by quantum theory. Going back to the story of our twins, what Bell proposed is to examine how the twins reacted in cases where the vending machines made different recommendations. If we assume that the twins coordinated their responses each morning before going to school, in accordance with Einstein's approach, Bell showed that in such a case we should see certain probabilities for the twins' reactions to different recommendations. That is, if we assume that the principle of locality is valid and there is coordination in advance, we get probabilities of a certain kind. If we could examine these probabilities experimentally and they match Bell's expectations then we could prove Einstein's claims that coordination in advance has taken place, and hence that quantum theory still lacks the parameters which could explain this coordination. However, if we prove that the probabilities are different than what Bell expected it would mean that there was no coordination and that the twins managed to keep each other up to date in some immediate way. That is, there are two possibilities: the experiment's results confirm the "Bell inequality" and therefore quantum theory's current predictions are wrong and should be replaced with a new theory; or the experiment's results will disprove "Bell's inequality", and thus prove quantum theory's predictions and their implications regarding the EPR paradox to be correct, perhaps even disproving the assumption of locality.

"Bell's theorem" paved the way to experiments which would prove or disprove the implications which follow from the phenomenon of quantum entanglement. But Bell's theorem has another important implication: it assumes the property of locality in the world, and so if it is disproved then so is the assumption that the world is local, and any hidden variables theory that follows will have to be a non-local theory. If it is proved that the theorem is correct then that would mean that quantum theory doesn't contradict itself, and then the non-locality of quantum entanglement must be explained. So in any case there is a big question mark regarding the property of locality. But maybe we should be asking another question: is it possible that the correlation isn't related to an actual transfer of information between the particles, and that we need to find another explanation? Einstein, Podolsky and Rosen, proposed two possibilities: either the correlation exists in advance or we need to find an explanation for it in the theory, or there is a transfer of information which is faster than the speed of light. Could there be a third explanation for this correlation? We will expand on this later on. It is also important to understand that Bell's theorem does not contradict the possibility of the existence of "hidden variables" theories, but it does bind them to non-locality.

At this point it's finally possible to examine if Einstein's thought experiment, which was meant to disprove quantum theory, has found a flaw in the theory itself, or whether it has identified a violation in nature of one of the most fundamental properties we know of (locality).

Only in 1982, about 50 years after the publication of Einstein, Podolsky and Rosen's article, was the technology developed with which an experiment meant to examine "Bell's theorem" was performed. Alain Aspec, a Frenchman, carried out an experiment with two photons in a state of quantum entanglement and the result was unequivocal: "Bell's inequality" was violated. What is the meaning of this violation? The formulation of "Bell's inequality" which Aspec examined in his experiment was based on two assumptions: the assumption of objectivity and the assumption of locality. The assumption of objectivity means that the objects we talk about, and their properties, exist in reality itself, and their existence doesn't depend on the existence of an observer. Put simply, there is a world out there that has an existence of its own. The assumption of locality means, as we've seen, that an immediate effect of one object on another in nature is not possible and that relativity theory sets a limit for such an effect - the speed of light. Aspec's experiment violated the inequality and so disproved one of the two assumptions: either there is no objective world, or the world is non-local. The two possibilities aren't easy to stomach, however it is surely easier for us to accept an objective world which has non-local effects over a world that doesn't exist objectively and that has local effects. For this reason, most physicists today accept the claim that Aspec's experiment proved that effects at a distance exist, and that the property of locality does not exist in nature.

The meaning of this is huge and inconceivable; the most fundamental components of the world affect one another in some way, at a distance, without ever transferring information between them or interacting physically. To this day it's not really clear how this happens, but since Aspec's experiment the phenomenon of quantum entanglement was examined in many experiments and is nowadays considered to be a scientific fact. Furthermore, use of it has become a common thing in physical research and there are even first stages of progress towards practical use of this phenomenon, mostly in the world of quantum computing and cryptography (the study of codes and their decryption).

Even though quantum entanglement is more reminiscent of feats of magic than a physical theory, we can try and understand its implications. I will try to present the different possibilities which follow from the attempt to understand its implications on our world and perhaps also the concept of time. But before we delve into the concrete implications of this phenomenon, it is important to point out that it is possible that one day it will become clear that Bell's inequality is inexact, and that it will be corrected and reformulated in a way that would make it possible for quantum theory and the property of locality to exist side by side. Many attempts at correcting the inequality have been made, unsuccessfully so far. However, until it is proven otherwise, we cannot rule out this possibility entirely.

One of the possible solutions to the problem of quantum entanglement suggests that it's possible that the view regarding the arrow of time and the transfer of information from the past to the future is wrong. As we've seen in the previous chapter, the the symmetrical arrow of time view can lead us to the solution of controversial issues which follow from quantum theory: Hugh Price is convinced that the Bell's theorem's issue of locality can be solved by ruling out the assumption that the future depends on the past, while the past does not depend on the future. That is, according to him, we must assume that the particles' values depend on a future measurement **which hasn't taken place yet.** This assumption prevents the theory's conclusion of non-locality. Price claims that even though we expect causality to go in one direction, we can expect symmetrical causality in time, so why shouldn't we assume that the future can affect the past just as much as the past affects the future? For this reason, he suggests that we view quantum entanglement as a phenomenon which demonstrates reverse-causality. According to this approach, the measurement of one particle in the entagled system, affects the other particle backwards in time and causes it to be coordinated with the results of the first particle's measurement in the past, before the measurement was performed. The correlation created is the result of a causal affect operating backwards in time, and so we can't explain it using a uni-directional arrow of time. Causality is bi-directional and the future measurement affects the particle's past values.

Another possible solution to the question of entanglement is the "common past hypothesis" which Bell discussed in his work. According to the common past hypothesis, the correlation between the two particles resulted from some past connection; that is, the correlation was not created at the moment of measurement but rather takes place in some way at the start of the connection between the particles. Bell himself rejected this possibility, mostly because it results in a world that's inter-connected in a "complicated conspiracy". In my opinion, it is hard to see how such a correlation could exist without us needing another "hidden variables" theory, which we already know would have to be non-local anyway.

Another interesting possibility, in my opinion, is that the two particles aren't really separate and are parts of one complete, complex system. Our observation of the world tends to be mechanistic: we see in the world a complex system made up of separate parts, which has to be taken apart if we are to understand it, according to the reductionist approach. The EPR Paradox examines every particle separately and tries to understand the entire system through a reductionist analysis of each individual part. However, we could also adopt a point of view which isn't mechanistic and is more similar to the views of eastern philosophy, according to which the world is one complete system that has no separate components, but rather each individual component is an inseparable part of a bigger picture. According to this view, we must look at the two particles as one system, and not try and take it apart. When we look at the system this way, it is possible that the property of locality doesn't exist in the sense of transfer of information at a speed faster than the speed of light, but that there is a connection between faraway components in the universe that are part of the same setup. The two unified particles are part of a continuous space or network, and the fact that they're part of one large system creates the illusion of there being a transfer of information between them. Through some means, a measurement of part of the system provides new information regarding its other parts. In order to better understand this possibility, think of a very long stick, say a stick that's a light year long - that is, transfer of light from one end to the other takes a whole year. If we wanted to transfer information between the two ends of the stick at highest possible speed it would take us a whole year. You're standing on Earth and holding one end of the stick. If you know the length of the stick and angle at which it is being held, you could calculate, fairly simply, where its other end is located. Now you move the stick slightly. This small change you've made here on Earth will have a huge influence of the location of the other end. Theoretically, the information regarding the location of the other end is unknown to us, and in order to receive it we would have to wait a whole year for it to arrive here on Earth. But in practice, any mathematician would be able to calculate the exact location of the other end according to the change in the position of the end that's located on Earth. That is, the information regarding the location of the other end will be known to us long before it comes to us travelling in the speed of light, by virtue of the fact that we're talking about one system which includes both ends of the stick. This is of course a simplistic analogy which doesn't come close to being a satisfactory explanation for the coordination between the two particles, and so we'll make things a little more complicated: imagine a network of points scattered over different locations in the universe, which are far away from one another. Any change which takes place at one point of the network will affect the entire network. In this case as well, we wouldn't have to wait for all the information about the new positions of each point in the network to arrive to us in the speed of light, we could immediately know the new location of each point due to the fact that it is one, unified system.

Although these examples describe very simple systems which include two ends or a network of points with locations in space, I use them to raise the possibility that a view which sees the universe itself as a very complex network, which includes everything we know in the universe as one overall system, could perhaps begin to explain the coordination which takes place in quantum entanglement. This network-like model which makes it possible to know locations in the network, for example, using other points in the network without an immediate transfer of information, solves the problem of locality and does not contradict relativity theory, and thus solves the EPR Paradox. It is possible that quantum mechanics forces us to see the world not as a collection of separate physical objects, but as a complex network of relations between different parts of the one unified whole.

It is hard for us to even imagine a clear visual picture of such a network. How is it connected? What is it made of? How can the wide and varied collection of properties we see in the various objects in the universe be part of one inter-connected network? These are difficult and inconceivable questions at this stage. However, imagine you go back to the 19th century and tell the people of the time about electronic devices which are connected to one another in a world wide network through which you can transfer any kind of information - an image, a video, text or any other thing you can think of - and share it with all of mankind. A 19th century man wouldn't be able to begin to understand what you're talking about when you try and describe the network that is the internet to him. Would someone from the 22nd century be able to convince us, people of the 21st century, that the "network of the universe" is the foundation of everything we see around us?

I claim that the phenomenon of quantum entanglement, and many other phenomena related to quantum theory, should be explained using a network view of the universe. This is a highly philosophical approach at this point and I don't have the ability of expressing it through clear physical means. In order for us to understand it and construct the beginning of such an explanation I will describe in the following chapters the wondrous world of network theory as an example of the fascinating rules which follow from complexity theory. This theory is developing rapidly since it appeared in the end of the 20th century in the world of biology and is gradually penetrating every aspect of our life. In my opinion, physics will be the next significant step in the innovative and revolutionary influence of complexity theory on the way in which we understand the physical world. Later on in the book I will attempt to link between the big questions - including quantum entanglement and the concept of time - to the world of networks and complexity theory, and propose a new view of a reality built of a complex and inter-connected network whose components are in a state of constant interaction between each other, in which quantum entanglement is the cause for the creation of each and every moment in the life of the universe.

Since the Aspec experiments in the 80s, quantum entanglement has been a very significant part of physical discourse, and there are those who believe that it is the foundation for a much deeper understanding of the universe. One of the most fascinating theoretical implications of quantum entanglement, which in later years has received experimental reinforcement, is the possibility that this phenomenon explains the emergence (formation) of space and time in the universe. If this possibility is true, then quantum entanglement will be able to answer the hard question that have remained open regarding time. But this is where I'm in need of your patience: in order for us to be able to discuss these matters in a clear and simple way and link them to everything we've dealt with so far, I must stall a little longer and discuss the understanding of complexity and network theory, and before that, to stop and understand the difficult problems we encounter when we attempt to understand the concept of time in the intersection between quantum theory and relativity theory.

CHAPTER 9:

The Death Blow - The Concept of Time in the 20th Century

The journey we're going through here has a destination. Every step we take and site we stop at are meant to eventually lead us to the right point at the end of the trip. It is a "treasure hunt", where each point includes additional knowledge, a new piece of information which will allow us to find the treasure. And that treasure is - rescuing the concept of time. "Time" as we know it today, in philosophy and in science, is in need of rescue. It is difficult not to arrive at the conclusion that time is in a state of advanced death throes, and many physicists and philosophers have announced its death a long time ago. The end of the concept of time is basically the understanding that what we experience as time, as a constant and consistent progression of change and movement in the world, is merely an illusion. The goal we set up for ourselves is a pretentious one, to say the least. 2,500 years of philosophizing and acquiring knowledge about the world have led mankind time and time again to the conclusion that time does not exist and is merely an illusion. Every generation sees attempts to breathe life into the ontological concept of time and to prove that it exists in the world, and each time it loses the physical foundations on which it moves and freezes like a statue when it comes into the terrifying gaze of the white witch - physics. Since I do not intend to go down the same dead-end path which many have taken before me, I will try to break the pattern and offer an explanation of the concept of time from a fresh, innovative point of view. I will not Sisyphically roll the rock to the top of the mountain just to find it at its bottom the next day, but rather I will leave the rock unmoving, in the place where it stands today, at the bottom of the mountain, and attempt to breathe life into it. In order to understand the metaphor and the explanation I'm proposing, we will use this chapter in order to stop for a moment, sort out the information we've gathered up until now on our journey and summarize everything we know about the concept of time in 20th century physics and philosophy. In the following chapters we will turn to new worlds which have been developed mostly in recent years, and which I believe can constitute the foundation for the entire explanation - an explanation I will attempt to flesh out in the book's final chapters.

Before, Now, or Always? - "The Growing Universe", Presentism and Eternalism

Every point of view regarding the concept of time which exists today can be classified into one of the views existing in our philosophical and physical language: time as an existing entity or time as an illusion. Each such view can include components of time which can be ascribed existence in the world or components which cannot be ascribed existence, since they're supposed to exist in the future or they have already existed in the past and have passed from this world. Generally, there are two such central philosophical views of time: eternalism and presentism - the same old views I described at the start of our journey - and a middle view called possibilism.

The intuitive view most of us hold is the view called "presentism" (perception of the present). This view ascribes existence to the present alone, that is, according to it the only thing to which we can ascribe existence in reality is the present moment we are experiencing. Any past event has already disappeared and so has lost its existence, and any future event hasn't happened yet and so it hasn't acquired the property of existence yet. According to this view, you and the book you're holding can be considered to exist in this moment, however Newton and the quill he used to write the Principia cannot be considered to exist at this moment.

Any other view of time which is not presentism grants existence to other components beyond the "now". One such view, which also suits the intuitive view some of us hold, ascribes existence both to the present moment and to everything that has already happened, that is, the past. According to this view, anything that has happened up until now has acquired the property of existence by virtue of taking place in reality, and anything that will happen does not exist yet, instead it merely has the potential of existing. For this reason, this view is called "possibilism", or the "Growing Universe Theory". According to this view, you and the book you're holding, as well as Newton and his quill already exist, while the Andromedan who will visit the Earth in 2253 only has the potential to exist. This view allows us to keep the future as an open world that does not exist in advance, in which each moment is created anew. The past and the present already exist and are fixed, and the future is "the land of limitless possibilities".

The final view of the concept of time, which we've already encountered in previous chapters - for example in the "block universe" view - is eternalism. This template ascribes existence to all moments in the past, present and future. All of these exist in parallel and there is no difference between the existence of any event, person or object at any point in the history of the universe, as well as its future. While we only experience the present, this means nothing in terms of granting existence to any event which does not belong to the present moment.

It should be emphasized that the views which aren't presentism deal, of course, with a non-temporal-epistemological universe - that is, they don't deal with what exists in the sense of the point of time we're experiencing and perceive in the endless continuum of points in time, but with the ontological existence of the reality which lies beyond our perception. According to the eternalist view, for example, the fact that we feel the present moment as existing and the future as not existing is a result of our temporal and perceptual experience alone. It does not contradict the conclusion arrived at by the eternalist view, according to which, in the "real" world, all the events in the future already exist in the same measure of certainty and have the same type of existence as any other event in my past or present.

Now that we've described the possible world views regarding the concept of time, we will attempt to understand how physics and philosophy have led, step by step, to the eternalist view, and have thus killed the concept of dynamic time. In contrast to dynamic time, which is always passing, the one we all experience, they presented us with an image of a static universe, unchanging and unmoving. The long process which has led the world to the realization that the universe is static is mostly based on three milestones which have contributed to the understanding that the universe is static during the 20th century: the McTaggart argument, relativity theory, and the attempt to combine quantum theory and relativity theory into one equation representing the entire universe.

McTaggart's Axe - The Philosophical Argument which Struck at the Concept of Time

Throughout our journey, we've encountered quite a few philosophers from various periods in the history of mankind who have studied the nature of time, but one of the most powerful and structured arguments published in the framework of the debate on the concept of time only appeared in 1908 (unrelated to the publication of the special theory of relativity - 1905). John McTaggart, an English philosopher, published his argument, known today as the "McTaggart argument", in his book "The Unreality of Time". McTaggart focused on the philosophical aspect of the question of time and its ontology, and in effect demonstrates that any attempt to ascribe a property of change to time which is required in order to grant meaning to time is futile.

And this is how he constructs his argument: every event in the history of the universe can be arranged in a certain order, which we will call a "Series". A series is a continuous template of positions, each with a certain title - like a long train with a sign on every carriage which bears the name of that carriage. In each such carriage, we can place a moment in the history of the universe. Series A is a series of events in which each event bears the title of "past", "present" and "future". Each carriage in the train marks a certain point in time on the axis of time which includes the past, the present and the future, and every event in the history of the universe is assigned to one of the carriages. For example, event A is in the carriage which has the sign "two days ago" on it, and event B is in the carriage which has the sign "three hours from now". So, series A is the intuitive and familiar continuum of events on the axis of time. This is one kind of arrangement.

series B is arranged differently. It is a series of all the events in the universe in which each event is labeled according to its position relative to the others, that is, every event is described using labels of "before" and "after". The train's carriages are arranged according to the order in which the events took place and the sign on each carriage indicates its position, before or after another event. For example, event A is in the carriage labeled "two days before event B" or "three hours after event C". In this case the arrangement is relative to other events and does not reflect the absolute axis of past, present and future.

Note that the central difference between the two arrangements stems from the fact that in series A the events are placed continuously on an objective axis of time, with a past, a present and a future, and their positions change as the axis of time progresses. In other words, at each moment events are added to each carriage according to the progression on the axis of time. The events in series B, however, are not placed on an axis of time but are labeled only in relation to other events, and their position is static and unchanging. Each such event always stays in the same carriage, since event A will always come before event B (or after it, or at the same time as it) and this relation cannot change. However, the label of a certain event in the "future", as in series A, keeps changing since it becomes the present and the past and later on, the distant past. And so, in series A, each moment and each event are constantly changing their position and are moving to the carriage which is appropriate to the point in time which is relevant for them on the axis of time.

In order to take his argument further, McTaggart uses the conventional assumption that time has no meaning without change. We've already encountered this argument in different places and we've seen that time is essentially an expression of change; if we take change away from the world then the existence of time has no meaning.

The assumptions in McTaggart's argument, therefore, are:

1. There are two possibilities of arranging the events in the universe in a certain order.
2. One possibility is called "series A" and it is an arrangement based on the axis of time, with a past, a present, and a future. Each event receives an objective position on the absolute axis of time.
3. A second possibility is called "series B" and it is a relative arrangement of events, one after the other. Each event is positioned only in relation to other events.
4. In order to show that time exists in one of these series, we must find the component of change, without which time has no existence.

Now that we've presented the assumptions, we will move on to the argument itself. McTaggart claims that viewing time as a continuum of events in the framework of series B does not include the concept of change, since series B is fixed and static. Every event will always be in its relative position compared to the other events; each event will always be in the exact same carriage in this "train of events". For example, event A, which is described as "two days after event B" or "three hours before event C" will always remain in its position in the series, unchanging. Series B is a dusty pile of events arranged one after the other, and so it is a static image without the expression of change which is needed to prove the existence of time. In order for us to be able to use series B to describe events in time we need series A, which provides the dynamic component to the continuum of events.

While McTaggart does find the required component of change needed to describe time in series A, since every event isn't static, the title it receives in the series keeps changing. An event in the future which bore the title "two days from now" will receive the title "in a day from now" one day later, followed by "the present" and then "yesterday", "two days ago" and so on. That is, series A satisfies the required condition of change which every event goes through, the one series B doesn't.

The problem is that in series A, McTaggart identifies that each event eventually includes all possible properties - past, present and future. Each carriage in the train will "host" every event in the history of the universe since the labels are constantly changing, and each event will eventually bear every possible label of past, present and future. That is, every event can be described using labels which contradict one another and cannot exist simultaneously. An event cannot be both "past" and "future", in which case - what is the meaning of placing events in a certain order if we can find all of the events in each place in the series?

In order to solve the problem of simultaneous existence of properties which annul one another in each event in each position in the series, we can argue that each such property is given to an event at another point in time, that is, each event is "future" in a particular point in the past, and is "past" in a certain point in the future, and so these properties can never exist simultaneously. McTaggart mentions this answer but immediately rules it out based on a well-known argument: such an answer requires an additional axis of time, in addition to the axis of time we're trying to explain. As we've seen in the discussion on the flow of time, the need to classify events as past, present and future, and to give each such property a different position on the axis of time leads to a description of two parallel axes, and any attempt to solve the problem creates an additional axis of time. When we say that event A is labeled as "in two days from now", and 3 days later it's labeled as "yesterday" in time Y, we're essentially explaining the positions on the axis of time using an additional axis of time, which we will need to explain as well, and so on. This is precisely the problem of infinite regression which leads the argument to arrive at an internal contradiction. In other words, an attempt to ascribe existence to time based on series A leads us to a logical contradiction.

McTaggart summarizes: since series B is completely static and meaningless in terms of the existence of the concept of time, and series A contains an internal contradiction, there is no way of describing the existence of the continuum of time in a logical manner and so time does not exist.

This argument brings us back to various philosophical and theological statements, from Thomas Aquinas, through St. Augustine to Parmenides, and similar eastern views which view reality as a unity of all times. However, McTaggart was the first one to formulate an ordered philosophical argument based on the laws of logic, in which, based on clear assumptions we arrive at an ontological conclusion regarding the lack of a property of existence in the concept of time.

There are philosophers who subscribe to "B Theory": they accept the contradiction inherent in series A, but believe we should reject the conclusion regarding time's non-existence in series B. According to them, this world only has a description of time according to series B; that is, the temporality of events is relative, and they are not placed on a change axis of time with a past, present and future. This approach does provide time with existence to an extent, but it also leads us to the understanding that time does not progress and that the sense of the passage of time is an illusion of ours - or that it lies in other ontological components we do not yet know about. Prima facie, it's hard for me to see how this approach saves us from the predicament we find ourselves in.

Support for McTaggart's argument comes from the special theory of relativity. According to this theory there is no true simultaneity between events in the universe, and time is relative to the observer. That is, an event that for one observer is taking place in the present will exist in the past for another observer and in the future for yet another one. If so, arguments which use descriptions of past, present and future, cannot be true or false in an objective sense, and so the passage of time cannot be an objective property of reality.

The static theory of the block universe as arises from the special theory of relativity does indeed rule out the reality of the concepts of past, present and future (series A), but the order of "before" and "after" (series B) still remains.

Let's go back and summarize the part which the special theory of relativity played in turning our universe into a static one, and how it also ushered the end of time.

Einstein's Gun - The Loss of Simultaneity and the Concept of Time in Relativity Theory

Now that we understand the philosophical argument, which is the first factor among three which, in my opinion, led the philosophical and physical communities to sink into the bleakness of the loss of time, it is time to understand the most significant physical argument when it comes to removing time from the universe. To summarize everything we've said about the special theory of relativity and its implications regarding the perception of physical time in the universe I will include here a summary of the central messages which are relevant to our discussion, and will reinforce the insights which stem mostly from the special theory of relativity regarding the disappearance of time from the physical picture.

I remind you, that the constant speed of light has led, as part of the framework of Einstein's special theory of relativity, to the understanding that time and space are relative to the observer and that they contract and stretch in accordance with the speed and direction of the observer's movement. The direct meaning of relative space and time is that there is no "privileged" observer, whose present is the universe's present, and to which all other observers are relative. Since we don't have any such privileged observer, each observer's point of reference when it comes to space and time has its own relative truthfulness for that observer. Put simply, none of us is privileged over the other and so there is no time which can be defined as "this moment" throughout the entire universe. My present isn't the same as the present of my Martian neighbor. There is no one configuration which describes all the events in the universe and the positions of all the objects in the universe, which can be pointed at in a given moment and defined as the absolute present, to which the rest of the bodies in the universe are relative. In Newton's universe we can "stop" the universe at a certain moment and receive a description of all the events taking place in the universe at that moment for anyone in it, without the velocity of one of those people affecting this picture. The special theory of relativity has shattered this convention. If so, if each of us has his own different configuration of events which are simultaneous at a given moment and these events are not the same simultaneous events for another observer at that moment, it follows that the universe has lost this property of simultaneity which was so fundamental to Newton's world view. The events which take place simultaneously in the universe, in my current configuration, are not identical to events taking place simultaneously in your configuration at this moment. This is the meaning of the loss of simultaneity in the universe.

And yet, why does the loss of simultaneity imply the loss of the concept of time and the rise of the "block universe"?

The next move is a controversial one, mostly since its significance is inconceivable to us as human beings operating freely in the world. Up until now, all the attempts to topple the foundations on which the block universe argument was built have failed, and the tendency of most people working in the field is to rule out the idea entirely, setting it aside and going on with life as usual. This may be the right thing to do psychologically speaking, but with your permission, since I'm not concerned with the psychological implications of this book, I will proceed to establish this argument in the following paragraphs.

The argument I will present here, constructed by philosophers and physicists, attempts to explain the block universe based on the loss of simultaneity in the world. Firstly, we will begin with the assumption that what goes on in our world has existence. Thus, for example, at this moment I'm typing these words into my computer – this is a moment which exists. I cannot know if any other event, which does not exist in the present, has existence - maybe it lost its existence or hasn't gotten it yet - but regarding this event I can know for certain that it exists. Secondly, according to the special theory of relativity, every two events which take place simultaneously from the point of view of one observer have identical existence as far as he's concerned. In the argument I'm constructing here I will use these assumptions to examine the existence of all the events beyond the present. In order to understand the next move in the argument, we must familiarize ourselves with one of logic's most basic rules. This rule states simply that if A=B and B=C then, certainly, A=C.

Now, we take three events in the history of the universe: A, B, and X. According to the logical rule we just learned, if event A and event X have equal existence at this moment and if event B and event X also have equal existence at this moment, then we can conclude with certainty that event A and event B have equal existence in this moment.

And now for the coup de grace: we define events A and B as events which are distant from one another in time. Event A is the moment in which I write the word "this" and event B is the moment in which an asteroid hits the Earth in a few tens of thousands of years from now and wipes out all life on it. According to the total loss of simultaneity in the universe which follows from relativity theory, there will always be an observer, Bob, for whom event A will be simultaneous with some event X, and there will always be another observer, Alice, for whom event B will be simultaneous with the same event X. For Bob events A and X are simultaneous and so have equal existence; for Alice events B and X are simultaneous and so have equal existence. And in accordance with the logical rule we learned, if A and X have equal existence and so do B and X, then A and B have equal existence. The problem is that events A and B take place tens of thousands of years apart, and yet they still have equal existence.

This move can be carried out for any event in the history of the universe, in its past, present and future, and so all the events have equal existence. This is the "block universe" argument: the block universe is a static universe in which all events already exist without a future being created out of the present. Our experience when transitioning in this four-dimensional block of events, in the way in which time is constantly being created, is merely an illusion. If the universe is a static block of events then dynamic, gradually progressing time has no physical meaning, and so it must be taken of out of the physical picture of the world.

And yet, even those who are familiar with this argument and have yet to find a counter-argument which disproves it, go to great lengths to ignore it, due to the controversial implications of this argument when it comes to our freedom of choice. If every event in our future already exists in this moment then any decision we make in the future has already been made, and the feeling that we're living in a world with free will and the ability to make choices is a complete illusion, or in other words - absolute determinism. A regular person cannot perceive our world in such a way and keep on living in it knowing for certain that this is our fate.

Over the years, many have claimed that the "block universe" argument does not explain the issue of the passage of time, which is so strong in the way we perceive the world, and the fact that there is a clear arrow of time in the universe. There are several types of arguments against the block universe: some of them claim that relativity theory is incomplete and that it cannot be fully applied to the universe; others claim that despite everything there is some privileged moment or absolute reference point in the universe according to which all observers should be placed, but we - and relativity theory - aren't able to identify them yet and separate them from the rest of the relative reference points; some arguments attempt to find the solution in quantum theory and in the many worlds theory (we will discuss this later on). Another interesting argument against the "block universe" seems radical but comes from one of the world's most famous living theoretical physicists, Lee Smolin. In his book "Time Reborn", he claims that the laws of physics themselves change over time; the attempt to apply physical theories, such as relativity theory or quantum theory as we know them today, on the entire universe throughout the entire duration of its existence is a reductionist move of what Smolin calls "doing physics in a box" - that is, constructing theories in a lab and applying them to the entire universe without any physical justification or the ability to test them empirically. We will expand on Smolin's position regarding the concept of time in later chapters.

The next step in the death of time came a little later, with the attempt to reconcile relativity theory with quantum theory.

Wheeler-DeWitt's Poisonous Pill - The "Cursed" Wave Equation of the Universe

At the end of the 19th century, Lord Kelvin announced the coming end of physical research and the arrival at the goal of understanding the entire universe. The two clouds that still remained, according to him, eventually turned things on their head and completely changed physics and the way in which we understood the world in the 20th century. At the beginning of the 21st century, we had two physical theories which were both complete and well founded, which were backed by in-depth empirical research and the ability to provide accurate predictions on the macroscopic and microscopic worlds that weren't controversial in any way. Quantum theory perfectly describes the way in which the microscopic world works, and relativity theory does the same for the macroscopic world. However, a heavy cloud still hovers over these two theories, one that hasn't dispersed yet, despite accompanying us since the day those theories appeared on the stage of science - the two theories contradict each other. Albert Einstein, one of the trail blazers who brought these two theories to the world, dedicated his best years to the attempt of dispersing this cloud, but to no avail. Since the start of the previous century until today, the world's best physicists are looking for "the theory of everything", the theory that would be able to explain both microscopic and macroscopic phenomena using a singular physical and mathematical language.

Relativity theory and quantum theory still do not live side by side in peace, mostly since today there is still no possibility of combining microscopic quantum phenomena with gravity as described in the general theory of relativity. Without getting into the problem's technical details, which are complex and not that important in the context of the questions we're trying to understand in this book, when we attempt to describe the entire universe using quantum theory, we run into a contradiction between it and relativity theory, a contradiction which no one has been able to solve to this day. The central problem stems from the understanding that space-time (which is described by relativity theory as a curved, uniform and smooth space) cannot be divided into fundamental units which can be described as quantum units, and the attempt to explain the effect of gravity on these sub-atomic units (described by quantum theory as separate components) currently leads to contradictions, as well as unacceptable mathematical results. The problem exists in any description of space-time's fundamental units but gets worse when we try to describe sub-atomic points in space-time which have a large mass, such as in the center of a black hole or the first seconds of the big bang. What's interesting for our purposes is that some of the attempts to solve the clash between relativity theory and quantum theory are directly related to the question of time. The large weight of the concept of time in the gap between these two theories and one unified theory which would provide us with a singular, complete description of reality only strengthens the insight with which we began our journey. This is a big clue to the fact that something in our understanding of the concept of time, which hasn't changed significantly in the past 2,500 years, has to change in order for us to able to announce the next revolution in our understanding of the physical universe.

In 1980 the physicist Karel Kuchar gave a lecture in a conference bearing the name "quantum gravity". He summarized his lecture by describing the problem in connecting the two theories, using words which left a strong impression on the physical community and which still echo to this day: "the problem lies in the way in which the polar opposite points of view of quantum mechanics and relativity theory view the concept of time."

When we attempt to connect between quantum mechanics' wave equation and relativity theory's gravity, we get a wave equation that does not include time as a variable. In the mid-60s, John Wheeler and Bryce DeWitt published the wave equation for the entire universe. The equation is supposed to connect relativity theory and quantum theory and provide us with a complete view of the universe as one quantum system which includes gravity's relativistic point of view. This famous equation is known to this day as the Wheeler-DeWitt equation, and it describes the probability of any configuration of the entire universe. However, DeWitt himself called it the "cursed equation", since the equation obtained from relativity theory and quantum mechanics doesn't include time as a variable. Since we're looking at the entire universe, we can't place a clock outside of the universe to measure the development of the system on the axis of time, and so the Wheeler-DeWitt equation, which describes the quantum state of the universe, ultimately leads to a universe that is entirely static, a universe with no development. This is where the conclusion that time does not exist comes from. That is, the combination of quantum theory and relativity theory leads to a world view of a collection of moments (configurations) which are frozen in time. Sounds familiar?

Make a note of the point I made in the previous paragraph regarding the clock that's external to the universe. The meaning is that when we examine how a system develops, we do it using the axis of time of a clock that's external to the system, and if we examine the entire universe, we don't have such a "divine" point of view in which to place a clock. This issue will accompany us later on.

The Wheeler-DeWitt equation has great physical and philosophical significance, but it of course leads to great controversy - mostly since it's not a complete theory that will bring about the next revolution. It is easy to attack this "cursed" equation so long as no other unified theory appears which includes the solutions of both quantum theory and relativity theory. Decades of intensive work by scientists all over the world aimed at constructing this unified theory have produced many theoretical possibilities and developments which are very interesting. Some of the solutions attempt to find the unification using points of view that are a little different from existing theories; other attempt to find a new, more fundamental theory, such as string theory, which will explain the world using a more primary layer of information from which quantum and relativistic insights can be deduced. And there are those who attempt to claim that one of these theories (relativity and quantum theory) aren't true and must be replaced. However, no complete solution has been offered yet. Later on in this book we will examine two modern points of view from recent years which examine opposite ways for dealing with the problem: on one hand Julian Barbour, who in his book "The End of Time", comes to terms with the static universe and welcomes it; and on the other, Lee Smolin, who in his book "Time Reborn", does not give up on the concept of time and the fundamental essence of the passage of time and change in the universe, and attempts to find the way to bring time back into the physical picture of the world. According to the names of these books you can already understand just how deep the rift the scientific community finds itself in today goes, and how close the points of view are to those of there predecessors from 2,500 years ago: Parmenides and Heraclitus.

***

Once again we're back in a timeless world which includes a large collection of possible configurations, each of which has a certain probability of being actualized in reality. But this world does not contain any change or motion, and so it does not contain the concept of time.

As humans, it is hard for us to deal with such a grim world view of a static universe which does not truly develop and does not allow us to act as independent agents with free will. Is there a way to breath new life into the universe and wake it from its winter slumber? Will we see the arrival of a hero who will fight the white witch and lift the curse?

The answer to these questions isn't clear and has been discussed constantly over the past years. However, do not despair - we are at the start of a process which arouses great optimism in those who're seeking to bring vitality back to the universe. I'm not sure just how aware the physical world is of the coincidence which led us to the fact that we might find the best way to explain the universe in terms of the concept of time. However, I am convinced that a new way has been paved, the combination of which with the scientific world views of the late 20th century and the early 21st century can lead to revolutionary move. From this point onwards my goal will be to begin constructing the story of the 21st century regarding the concept of time.

In order to begin talking about the next step in the development of the concept of time in physics and philosophy, we must first familiarize ourselves with a new world, one that is entirely different from everything we've talked about up until now, which has developed in the fields of biology, sociology and economics, of all things, but which in my view will become a central player in the fight against the wicked witch of the north. The next heroes joining our story in a storm are complexity theory and network theory.

CHAPTER 10:

Doctor Chaos and Mr. Order - Complexity Theory

"The Theory of almost everything" - Bringing Mystery back into Nature

Our world is filled with mystery and unsolved riddles. Despite the feeling that we know so much about the world, there are still countless questions and gaps of uncertainty in a wide array of fields in the physical universe, in human society and in our everyday lives.

Some of the mystery is very fundamental and far from being solved, with questions such as the relation between the mind and the body, the creation of consciousness and the way the human brain works, the origin of life and the transition from still matter to a living cell, and so on. Some of the mystery stems from the fact that we can't construct accurate and complete theories which will provide us with one clear answer, for example, to predict the weather, build templates and provide accurate predictions in economical markets and the stock exchange and the large amount of vagueness when it comes to psychotherapy. Even in our everyday life, we're surprised by processes which we know well in our personal, work, and community life, and it doesn't matter how much information is revealed to us and how much we try to plan, manage and review these processes. There has always been a tension and a gap between the theories which attempt to explain the world and our everyday experience and the failure to apply these theories in "real life". It seems that the theory is a sterile description made in laboratory conditions and that its application in everyday life never fails to surprise us, and eventually we almost always identify a large gap between it and the results in practice.

Uncertainty and lack of knowledge have always accompanied the human race. We've always been looking for answers to the mysteries that surround us and we've filled up that gap through faith, religions and scientific theories. At first, most answers were found through faith in idols and religions which developed in tandem in various places around the world, and later on this was combined with the attempt to understand nature through philosophical thought and modern science. The combination of science, philosophy, and religion created a hodge-podge in which scientific theories are constructed whose purpose is to remove the veil of mystery from the world, while keeping God and faith in places where the gaps in our understanding of the world remained. Many people of science combined scientific theories with theological and mystical views, and one of the most well-known examples is of course Newton himself, who was a very religious man and spent a great deal of time studying theology and mysticism. Even in his Principia, he left room for God to "reset" the planets' motion through the solar system every few years, using comets. Newton used God when the equations didn't fully match the predictions which accumulated over the years.

The problem we get when combining religion and science is known as the "Lord of the Gap". This problem was created since God's place in the physical and scientific picture was growing smaller and smaller, and eventually the desire to combine faith with scientific theory caused God to lose his stature and become just another theory which accompanies science in a parallel lane. From the point of view of the believer, instead of giving God his lofty and expansive position in the world without contradicting any scientific view, he became an alternative to which better substitutes have been found over time. The believer can always claim that God's place hasn't taken a hit in the scientific world view which developed over time and that he's always present in it, even in the processes which science has succeeded in explaining. However, this is a change in God's status: from an active factor which is the reason for observed phenomena, he's become a passive factor, who just needs to make sure that everything is going smoothly. A kind of watchmaker-God who's sprung his watch and let it run, only interfering very rarely to reset its course.

The process of lifting the veil of mystery from the world has led mankind to set up the goal of constructing the "theory of everything": one scientific theory which will explain the entire world, from microscopic building blocks to the macroscopic universe, through human consciousness and social processes. Many still believe that such a theory is within our reach, but I believe that the most important lesson we can learn from the 20th century and the start of the 21st century is that the universe and mankind essentially contain fundamental components of uncertainty and deep complexity, and so we will never have the ability to acquire the information required for an accurate prediction of the universe. The past few decades have taught us that the world is fundamentally and inherently unpredictable and is submerged in the murky waters of randomness and uncertainty. Three insights have established this understanding: the relativity of the observer, quantum uncertainty and chaos as a routine state.

We've already met relativity theory, which brought with it the relativistic component of nature and the human observer; this component was significantly reinforced by quantum theory, which added the most fundamental physical "uncertainty" to the universe. However, much to the dismay of anyone who still had faith that we can build a deterministic-Laplacian picture of the universe, the second half of the 20th century and the start of the 21st century added new insights into the mix that did away with any shred of such optimism. These insights are the worlds of chaos and complexity.

As you recall, the physical view up until the 20th century - and to a great extent, the view among a large portion of the physical community to this day - is mechanistic, that is, it expresses the view that the world operates like a machine. All that we need to do is take the machine apart, down to its components, and understand how each part works and what it does, and thus we can understand the entire system and predict how it will behave at any given moment in the future. This is precisely the view that led to the Laplacian insight, which claims that if we know the information regarding every single particle in the universe, we can predict what the world will look like at any moment.

The mechanistic world view was so strong that it was applied to almost any paradigm and field in our world: from physical and biological theories for understanding nature, through psychological and sociological views which explain social and personal processes, to the understanding of business and economic processes and structures. And yet, it seems that these theories deepen our world, but only describe it up to a certain point, and there's always something missing which prevents us from fully applying them and predicting the way the world will behave.

During the 20th century, something new and very interesting happened: the mechanistic view began to collapse. Out of the new physics (mostly quantum theory), the understanding that there are processes which include the component of uncertainty, and that nature has a probabilistic aspect, began to solidify. For this reason, we cannot truly know the system down to its smallest details and accurately predict what will happen in the next moment; not because of technical limitations that stem from a lack of information but due to the nature of the universe itself. In other words, systems in nature aren't so similar to the complicated machines that we're built since the industrial revolution after all, and they behave in a way that's unpredictable and different.

Another factor, which is even more significant in this context has to do with the growing understanding of biological systems. In the 20th century, the understanding that biological systems cannot be taken apart to their fundamental components and be understood using a reductionist approach, became clearer. Biological systems include properties which cannot be found in the fundamental components of matter and living creatures; life isn't in the building blocks of the living creature but in something which stems from a general observation of the complexity of the system itself.

Our journey has gone through a twist. We're embarking on a new path that will provide us with new tools and thought patterns that will allow us to better understand the mysteries of the world in general, and specifically, those of the concept of time.

The Universe Searches for Sympathy - Synchronization

One summer morning in 1666, Christiaan Huygens sat in his study, looking at the warm sunrays dancing on the curtain of his window, as he listened to the ticking of his clock. Huygens had already invented the pendulum clock and so enjoyed sitting and proudly listening to his two pendulum clocks ticking on his wall. Suddenly he felt peaceful and complete. The calm and connection he felt with the universe around him during those moments were familiar, but for a moment he couldn't explain this feeling. A few seconds later he realized what had changed in the room: the sense of unity and peace stemmed from the fact that the two pendulum clocks were suddenly ticking in unison, as though the two pendulums were following one another and coordinating their movement in a uniform dance of time.

Huygens, being a scientist, was curious with the phenomenon and decided to examine it. He interrupted the coordination between the pendulums and waited patiently. After a while, to his surprise, the two pendulums began moving with perfect coordination. Huygens examined this phenomenon again and again and published the results of his research among the scientific community, which he called "an odd kind of sympathy between clocks".

Huygens discovered one of the most fascinating phenomena in nature, which has countless different expressions: the phenomenon of synchronization.

The explanation for the phenomenon is relatively simple: at first, when the movement of the pendulums wasn't coordinated, some of the energy of movement of each pendulum passed through the walls and arrived at the second pendulum. Thus, different levels of oscillations (energy) passed from pendulum to pendulum through the walls and the gap between the oscillations caused a change in each pendulum's oscillation energy level. The process of energy transfer and the feedback loop of the new energies gradually led one another to more balanced energy and oscillation levels. The smaller the gap between their oscillations, the more the reciprocal feedback and influence decreased, and they finally arrived at the same point of balance and equilibrium of coordinated swinging. This is the explanation for the synchronization between the two pendulums. Similar processes have been observed in nature, such as the synchronization of the light produced by fireflies which has been observed along rivers in south east Asia as early as the end of the 19th century. Hundreds of thousands of fireflies along many kilometers turned their lights on and off in perfect coordination and created a flashing strip of light, which stunned the European discoverers who visited the place for the first time and could not explain the phenomenon. A less exotic example of synchronization can be found in the world of theatre, as soon as a play ends, when the crowd is clapping, an act which begins as a cacophonous noise of clapping at different and random paces, which, in seconds, becomes a coordinated cascade of clapping in an increasingly uniform pace. It's no surprise that there's coordination and synchronization between living individuals who transfer information to one another and respond accordingly until equilibrium is achieved, however what Huygens discovered was something much deeper, and the its full implications only began to seep into our consciousness in the 20th century. Processes of synchronization take place in physical, non-living components, and they are an expression of a much more fundamental process in the physical nature of the world.

Phenomena of synchronization in nature have been observed over the years but were not addressed scientifically, and the scientific community has treated it as an amusing curiosity and nothing else. Only in the 20th century, following the mathematical work of Norbert Wiener and Arthur Winfree, did we begin to understand the processes in which synchronization between elements which carry out actions which repeat themselves (oscillators).

Oscillators began to be discovered in different systems. For example, neurons in nervous systems fire electric pulses at different paces. When a collection of neurons begins working at a uniform and synchronized pace, the result could be an epileptic fit. In a study of earthquakes, it was discovered that pieces of tectonic plates functioned as oscillators and began pushing one another, building up tension, and when this level of tension crossed a certain threshold, the plates began moving and releasing the huge amount of energy which had built up in the form of an earthquake. One neuron firing or a small oscillation of one tectonic plate can create cascades of synchronization of all individuals making up the system, and thus cause earthquakes or epileptic fits. In every such system, a single element is under increasing pressure which builds up to a certain threshold, and after that it releases the pressure and transfers it on to others, creating a domino effect.

Arthur Winfree, a mathematician and biologist who studied biological clocks for years, including the rhythm of the heart, demonstrated in 1956 that there is a "phase transition" during synchronization. He showed that in a collection of oscillators there will be no synchronization until they've reached a certain threshold, and then a very quick synchronization will be created. The most central and relevant understanding in our journey in this context is that when a group of individuals operates in a random way and at different paces, a state of synchronization does not develop **gradually**. At a certain point in time, when the individuals pass a certain threshold of transfer of information between them (electrical, physical, sonic information, etc.), the system goes through a "phase transition" and "locks" itself in a state of synchronization. The best and most well-known example of a "phase transition" is the freezing of water. Water is in a liquid state of matter at room temperature, and even when it's cooled the molecules keep moving and water remains in the same state of matter. When the temperature arrives at zero, an immediate "phase transition" takes place which causes all the molecules to converge into a coordinated structure, and the water immediately goes from a liquid state of matter to a solid one. This rapid transition is an example of a "phase transition" which causes the synchronization of all the system's components and thus creates a new order in the entire system. Today, we understand that systems which are comprised of a large number of individuals which are operating randomly and influencing one another can, at a certain point, undergo a "phase transition" and rapidly create an order which did not exist previously in the system. We also know that such coordinated moves don't only happen in biological systems, but in fundamental physical systems in nature as well, and that these synchronized phenomena aren't caused by one intentional element operating with great accuracy in order to bring all the individuals to a state of synchronization - rather it is a phenomenon erupting from the joint and interactive activity of all the individuals in the group.

What does all this have to do with the question of time?

Two significant insights arise in this context. Firstly, we begin understanding that synchronization is a very fundamental phenomenon in nature. Synchronization appears in almost every system which is comprised of a collection of components working together. It is possible that the phenomenon of synchronization should have an expression in the primal theory of the universe, and perhaps the question of the origin of synchronization and the way it affects processes in the universe should accompany us in our attempt to understand how the universe works on the most basic level.

Aside from that, there's something about the phenomenon of synchronization that makes me treat it with an increasing curiosity. The inner feeling that's created in us as humans when we encounter synchronization in nature is, in my opinion, much more meaningful than it may appear at first. We look at a flock of birds in the middle of a synchronized dance in the sky, listen to electronic music which consists of repetitive patterns or look at fractal drawings which form geometric shapes which repeat themselves from within themselves - all these seem to us at first as chaos, and then, in one moment, we experience them as simple, clear patterns, which are constantly repeating themselves. At that moment, we feel a sense of peace and deep connection which is reminiscent of meditative feelings described in eastern mysticism, which searched for our unity, as well as the unity of nature. There's something about the phenomenon of synchronization that, in my opinion, creates in us a feeling that coordination and unity are a fundamental property of nature. The connection between such biological and physical phenomena and the primal and deep influence on the way we experience the world is a phenomenon which should be paid attention to and understood thoroughly as part of our attempt to understand the mystery of the universe, and of time.

The bottom line is, the process of synchronization has a unique and interesting result: order being created out of disorder. Chaotic systems suddenly acquire a regularity, as well as organized patterns of operation. How, in a world where entropy and disorder are supposed to constantly be growing, as part of the arrow of time we described in previous chapters, are organized systems created, in which order keeps increasing? This question will lead us to the next revolution.

The Stormy Butterfly - Chaos Theory

At the end of the 19th century, Oscar II, the king of Sweden, hosted a scientific competition, and invited physicists from all over Europe to submit their solutions to a variety of problems. The first problem presented to them was a disquieting problem which had accompanied the physical community since Newton's day, but for which no satisfactory solution was ever found, known as "the three-body problem". Henry Poincaré, a talented physicist who was active at the time, submitted his solution to the problem as part of the competition and even won it, although it eventually turned out that his solution was wrong. Poincaré's work paved the way for a scientific field called "Chaos". At the end of the 20th century and in the early 21st century, this field will act as the seed from which the next perceptual revolution, that of network and complexity theory, will sprout.

The three-body problem is a surprising one. We all "know" that since the days of Newton, science can explain with great precision the movement of the planets in our solar system and the effect of the gravity that's created between the different stars, however this impression has no solid foundation. According to the three-body problem, we can't accurately describe the movement of just three bodies, even mathematically. In the "Principia", Newton did a good job explaining the movement of two bodies operating in the framework of the attractive force between them, but when people tried to apply Newton's physics and mathematics in order to accurately predict how three bodies influencing each other would move, a discrepancy between the theoretical prediction and observations in practice was discovered. The fact that even in the case of such a simple system consisting of just three elements, it is difficult to provide an accurate mathematical and physical description has kept a lot of people busy and damaged their confidence in being able to describe the world with great precision. Many years would pass before we received an answer to this question. Without knowing it, everyone was waiting for a new and important tool to enter our lives, one that would allow us to do things we never thought were possible: the computer.

In 1961, Edward Lorenz, an American meteorologist, was working on his models for weather prediction. Despite complex and ingenuous mathematical models, the science of meteorology never succeeded in producing accurate predictions and was always criticized by the general public and the scientific community, due to the discrepancies between predictions and the fickle and surprising weather we're all familiar with. Meteorological systems are very complex systems with a huge number of parameters which can influence the final outcome. However, much like other systems, and according to the conventional way of doing things, if we focus on the most important parameters and use average values, we can obtain reasonable results which are close enough to reality. Lorenz made use of computers to analyze his meteorological models, and one day he came across strange results. He entered the known parameters into his computer and began receiving the predictions for the development of the meteorological system in the coming days. At some point he stopped the program and decided to re-run a certain time-segment. He took the temporary results which were obtained at the start of the segment and re-entered them into the computer in order to run the program again. To his surprise, on the second run, he received significantly different results from the ones he received in the first run. The data were the same data, the time was the same and the program was the same program. If so, Lorenz tried to understand, what made the weather prediction change so drastically? When Lorenz examined his work process, he noticed that the only difference between the two runs was that the numbers he entered into the system during the second time were rounded up to three places after the decimal point. So, for example, if the system gave a certain parameter the value 0.56789, during the second run it was rounded to 0.568 and re-entered in the system. The difference between the parameters was miniscule and almost insignificant, however when it lasted over time it had a huge impact on the final result.

Lorenz' significant discovery was that in the study of the weather, the effect of marginal, minimal values of parameters could be enormous, and so the meteorological system is a system which is sensitive to its initial conditions. What's even more interesting is the discovery that we can never provide accurate predictions, and that any predictions for longer than a week in advance are entirely useless. The reason for this is the well known "butterfly effect". Originally, Lorenz used the analogy of a seagull, however in 1971 he named one of his lectures "The ability to predict: does the flap of a butterfly's wings in Brazil lead to a tornado in Texas?", and since then, the expression "butterfly effect" has caught on. According to the "butterfly effect", "small" effects have an exponential influence: any small change creates a bigger change directly after it, which creates yet another change. That is, the influence of a small effect isn't proportional to its size, but rather it quickly takes on astronomical dimensions compared to their initial data, and so changes which may seem marginal can, over time, have a huge influence. In the case of the weather, pressure changes, air turbulence and rain have an accumulative and exponential effect. But what was truly interesting is that we discovered that this didn't just apply to the weather.

The insight that in certain systems, small changes to the initial conditions can have a significant effect on the results over time was not entirely new. Maxwell even wrote, as early as 1873, that "effects with physical values which are too small to be taken into account by a finite being can create results with a much larger effect". However, it was only in the end of the 20th century that people began understanding the implications of this insight in almost every type of complex system in our lives, as well as the mathematical models with which we can analyze them. Humanity, as we said earlier, can't skip up the steps of this knowledge without the computing tools which only appeared in the middle of the previous century.

The study of systems which are sensitive to initial conditions (which was eventually called "chaos theory") became central to certain fields, mostly in computer science, biology and economics. Researchers began explaining phenomena characterized by uncertainty using tools from chaos theory, since as it turns out, in systems which include many components, there will always be an infinite number of parameters, and so we can't fully predict their behavior. This isn't just due to a lack of a technical ability to predict the behavior of complex systems, but also because complex systems are fundamentally and essentially unpredictable. Certain parameters can be infinitely large (for example, an infinite number of digits past the decimal point), and so it would require an infinite amount of time to get to know the data.

The term "chaos" as an explanation for dynamic systems which are sensitive to initial conditions was coined by the physicists Li and Yorke in order to symbolize the random and unpredictable components of such systems, however with time it was made clear that such systems are interesting not just because they're unpredictable.

The "sad" conclusion stemming from chaos theory is that uncertainty and difficulty in making predictions aren't merely technical but are rather essential. We can never know all the components which affect a system and its parameters to a degree of infinitely many digits past the decimal point, and so we can't completely predict the behavior of complex systems. Alongside quantum theory's component of uncertainty, this means the end of science's dream of a Newtonian and Laplacian universe which operates like a watch, in which all we have to do is take it apart to its components, get to know all its laws and data and predict where it will be at any later moment. Now it's clear that even the theoretical possibility of the existence of a "the theory of everything" is ruled out, since any computer would need an infinite amount of data and therefore an infinite amount of time to gather and process them. Laplacian determinism is gone for good.

Chaos theory's gladdening result is the deep understanding of complex systems. Inspired by it, mathematical, theoretical and philosophical tools began being developed, which allowed us to begin understanding the operation of complex systems. When we look at nature, we encounter order rather than chaos; we see organization and development of simple systems into complex ones, rather than things falling apart into disorder, or converging to an atrophied equilibrium. Up until now we couldn't overcome the contradiction between the arrow of time and Boltzmann's increasing entropy, and the phenomena we observe in nature, those of an incredible orderliness which is being created and developed at every moment. Chaos theory began producing the insights that with time would become the widespread view of complexity theory. Systems which are sensitive to initial conditions seemed at first like chaotic systems whose behavior can't be predicted. However, as the study of biological, economic and sociologic systems progressed, patterns of order began appearing in chaotic systems. That is, chaos theory wasn't just a description of how chaotic the world is anymore but began to describe a picture of an order being formed and created of itself, during the development of such chaotic systems. Computer programs which were run over and over again on chaotic systems began presenting patterns of regularity and order which could not be found in the system's components, but rather only when observing the system as a whole. The next step would be to understand how order is created in a chaotic world, in which entropy is supposed to constantly be increasing towards complete disorder, and an equilibrium of a dying universe.

Putting Things in Order - Systems with Emerging Order

A flock of birds, acting as one, carrying out a synchronized dance which scares predators away; a line of ants marching over dozens of meters, finding its way to food, bypassing obstacles, adapting itself to the conditions of the path and getting back to the nest in one piece; a living body which begins as a single cell and develops towards a body which consists of hundreds of complex systems which allow for the development and inheritance of properties to offspring; and a consciousness which emerges from a collection of neurons which are connected to one another - the world is filled with examples of systems which develop towards a higher level of order. They are characterized by facing changing conditions and they feature entirely new properties and phenomena, which weren't found in the framework of the components which make them up.

How can such complex and elaborate systems manage to maintain themselves and develop to higher levels of order without any planning or directing element? What causes a collection of individual components to create one, comprehensive system which operates in an ordered and organized way which develops to higher and more complex levels of order?

At the start of this book we discussed the arrow of time and the explanation which physics proposes for the fact that we see irreversible processes in nature. When answering this fundamental question, we presented a complete and elegant solution which is based on the second law of thermodynamics, Boltzmann's interpretation of increasing entropy in the universe and statistical mechanics. The bottom line is, the second law of thermodynamics describes a world in which disorder is consistently increasing, until it reaches a point of equilibrium, due to the high probability of the system reaching this state. However, in reality, we see a world around us with countless systems which develop towards order, and a transition to more complex stages of development. The best example is life. The systems of life we see around us completely contradict the second law of thermodynamics since order is constantly developing in it: from a simple set-up of a few cells, they become developed, complex and organized systems of organs and whole animals, which work in cooperation with one another in order to sustain the system itself, to develop it and make sure to pass-on information to their offspring. How is this description consistent with the thermodynamic principle which explains the arrow of time in the universe? In other words, how do processes which create order out of chaotic systems exist in a world where the second law of thermodynamics establishes the regularity of increasing entropy?

The second law of thermodynamics provides an elegant explanation for the question of the arrow of time, but it relies on a problematic condition which hasn't been given enough attention up until now, and which could possibly explain the gap between theory and reality. The second law of thermodynamics is only relevant when it comes to "closed systems". A "closed system" is a system which isn't affected by external factors, that is, all the processes taking place in it are internal processes between its components, without any other influence. An example for such a system is a cup of coffee which consists of water, milk, sugar, and coffee. When we look at the cup of coffee, we see a "closed" system in which the various components are interacting with one another, eventually leading to the realization of the second law of thermodynamics and the mixing of the various components until they reach a state of full equilibrium between them. If we place the cup of coffee in a fridge, then the "closed" system now includes the fridge itself, along with the cup of coffee and its components. In the same way, the solar system can be considered to be a "closed" system in which there's an interaction of gravity between its components. So the definition of the "closed system" we're examining is subjective and depends on context. And this is exactly where the problem lies: no system in the universe is truly a "closed system". Any system we define as closed is influenced by external factors from which the system can never fully be insulated.

We can examine a system as a "closed" system theoretically if we assume that the influence of those external factors is miniscule, if it even exists, and so has no significance when it comes to the final result. However, as we've just seen, complex systems are sensitive to influences which are considered "marginal", and for this reason can have very large implications which we couldn't even manage to ascribe to those minimal external influences. Minimal influences could be an interaction between the system's components with external components, such as chemical reactions by the coffee cup's components with matter in the air, however there are also influences which we can never neutralize such as the influence of gravity. Gravity works almost everywhere and its influence on the system cannot be neutralized; it is also impossible to insulate a system from the influence of neutrino particles, which come to us from the sun and pass through Earth and us and any other matter in the universe at huge quantities every second. Neutrino sensors buried deep in the south pole identify neutrino particles which penetrated the Earth from the north pole and which passed through the Earth all the way to its other side. Such examples teach us that any attempt to examine a system as a "closed" system is merely an approximation of reality rather than reality itself.

If so, our first problem, which is an essential one, is that there is no such thing as a truly "closed" system, without any interaction or external influence, and so the second law of thermodynamics basically describes a state which is only hypothetical, and which doesn't fully exist in reality. The more difficult problem is that most of the world's systems are systems which have a real interaction with their environment, that is, "open" systems whose relation to their environment isn't marginal or minimal. For this reason, we need a more complete explanation for the fact that we observe a development of order in the universe, rather than only an increase in disorder (entropy), as Boltzmann and Maxwell's explanation predicts. We now re-open the question of the arrow of time, since the reply given to this riddle was based on the second law of thermodynamics; if the latter doesn't describe a large part of the processes we see in nature, have we really found the solution to the riddle of the arrow of time?

This is exactly the question which Ilya Prigogine, a Nobel winning Belgian chemist and physicist, struggled with. He spent most of his life dealing with the understanding and analysis of "open" systems which aren't in equilibrium. Prigogine was one of the founders of complexity theory and even tried to construct the theory's foundations and the language that would accompany its use.

Prigogine discovered that in systems that at the time were considered to be chaotic systems, which naturally and in accordance with the second law of thermodynamics, were supposed to converge to a state of equilibrium, there are certain patterns of action which repeat themselves and an order emerging from their chaotic operation. That is, in complex systems which aren't in equilibrium there's a process in which order and patterns of action are created. The order which emerges in a complex system doesn't necessarily lead to an atrophied equilibrium without the possibility of development but quite the contrary, order is created out of a lack of the system's equilibrium. We all know what water looks like when it reaches a boil: a "chaotic" movement of gas bubbles moving in whirlpools inside the bubbling liquid. However, if we focus on the process taking place in the water we can identify, amidst the chaos, circular and ordered patterns of whirlpools, cooling and heating, in a state with a distinct pattern. Prigogine identified this phenomenon and began searching for similar patterns in additional complex systems, and to his surprise he found such patterns in countless phenomena.

The insight that in complex systems, order is created out of chaos is about to change the way in which we examine our world. It can be applied to more than just physical systems in nature, and that's exactly what we'll do. First, the spotlight was turned to biological systems, in which the development of order, in contradiction to the second law of thermodynamics is a very prominent phenomenon which was already known at the time. Prigogine and his followers understood that in this insight lies the secret to life and its development. Since then this insight has already been applied to complex economic systems such as commercial markets and the stock market, and in the study of the development of international economics, as well as in sociological systems. In fact, these insights can be applied to any process in our everyday life since our world isn't in a state of equilibrium, but rather its normal state is chaos and a lack of equilibrium. For this reason, any attempt to converge to equilibrium and total order comes with prolonged frustration, and any attempt to fight it is doomed to failure.

Life on the Edge of Chaos - How is Order Created in Complex Systems?

In order to understand the way a complex system works, we must first define what such a system is. Professor Melanie Mitchell from the university of Portland defines this in her book as follows: "a system in which large networks of components with no central control and simple rules of operation give rise to complex collective behavior, sophisticated information processing, and adaptation via learning or evolution". Another, more focused definition she provides summarizes it as follows: "a system that exhibits nontrivial emergent and self-organizing behaviors".

Mitchell identifies several properties which complex systems have in common:

Complex collective behavior: there are large systems comprised of components which operate according to basic law without central control or an overseeing body. There are systems in nature in which order is created at a very high level using the cooperation of several complex systems in a total array, which adapts itself to changes and faces crises. All this is done without any directing or overseeing element, which planned and constructed the system and regularly monitors its action. The human body is the ultimate example of this. Another example is a swarm of ants: from a supposedly random operation of unintelligent individuals, the swarm manages to create complex underground engineered structures which protect the entire community - the queen, her offspring and the food being stored - as well as creating very complex logistic arrays, dozens of meters long, whose purpose is the identification of food and bringing it back to the nest while facing predators and changing obstacles in nature.

Signaling and information processing: a basic condition for the development of order in a system is processes of communication and feedback in which information is transferred between components. Such communication can be vocal, electric, chemical, physical (such as oscillations in matter) or any other kind of communication which will cause the components to change their behavior towards one another and create synchronization in different levels of action.

Adaptation: these systems adapt themselves to their environment and change their behavior in order to survive and develop as part of a process of learning or evolutionary processes. Rigid systems which operate based on a structured plan without any capability of flexibility will collapse in times of crises; complex systems which develop order must be, in essence, flexible and adaptable to the changing situation. When a line of ants runs into a stream of water, for example, the fact that the ants begin running around randomly will lead to them eventually finding the way around, and a line of ants will be created once again, with the ants safely marching towards food and then back to the nest. The fact that the components are dynamic and can carry out a wide range of activities gives the complex system the flexibility required to face obstacles and changes.

Time and Emergence - Emergence and the Duration of Time

In the context of complex systems, we need to learn about an additional concept which is "emergence". This concept will have a very important role in the remainder of our journey, when we attempt to apply all the knowledge we've gathered on the concept of time. In the process of creating order in complex systems, new emergent properties appear which don't exist in the system's components.

The mechanistic view's classic reduction suggests that we take the system apart to its basic components and through them, explain the phenomena which appear at the more complex levels. For this reason, physics is supposed to explain chemistry, chemistry is supposed to explain biology and biology is supposed to explain consciousness and psychology. We need to distinguish between an ontological reduction and a descriptive reduction. An ontological reduction claims that any complex system ultimately consists of basic components, but there are phenomena in the system which can't be explained just by understanding these components. Rather, an understanding of the operation of the entire system is necessary. The human body, for example, consists of atoms and particles and has no additional basic component (spirit, life force, etc.), however, it features phenomena which can't be explained purely in terms of those particles. A descriptive reduction is a wider application of the principle of reduction, and it claims that the complex system and its phenomena can be fully explained using its basic components. According to the descriptive reductionist view, if we truly know the system's basic components, we can predict and explain any phenomenon that appears at a higher level of complexity. Thus, for example, consciousness or life can be explained using an advanced theory of particles, without needing anything else. Most modern thinkers would agree with the ontological reduction but not with the descriptive one. Regarding the human body, for example, most thinkers would agree that the body really does consist of nothing but particles, but an understanding of just the particles in a reductionist way won't let us explain every phenomenon in the complex body; for this reason, we need to understand the complexity itself at a higher level, using a comprehensive view of the entire system. The complex body will always display phenomena which don't appear at the particle level and they cannot be used to make prediction using a deeper understanding of the particle world.

In a glass of water, for example, we see emergent properties emerging from a large collection of molecules. The molecules which make up the water aren't liquid or wet, that is, we can't find the property of "liquidity" and "wetness" in these components. The water's "liquidity" and "wetness" are new phenomena which appear in the macroscopic world in an emergent way, resulting from the interaction of all of the system's components, which we call "water". "Heat" is also such a phenomenon. The particles which make up matter don't have a temperature. What we experience as heat or temperature is an emergent phenomenon which results from the velocity of motion of the particles which make up the material, which creates a phenomenon in the macroscopic world which we experience as temperature. Life too is an emergent phenomenon which results from a sophisticated system of cells. We can't find life in a cell, and life as we know it stems from a complex system of cells working together. Consciousness is an emergent phenomenon which results from the neuron network in the brain. You won't find a thought in a neuron, a thought appears when we connect a huge amount of neurons, who're physically and chemically interacting with one another. Trends in the stock market over time are also emergent: they appear out of a collection of purchases and sales of the sum of individuals in the market. There are countless more examples in nature, the social world and the human world.

One of the greatest philosophers of the early 20th century, the Frenchman Henri Bergson, identified the uniqueness of biology with an additional force he called Élan Vital. Bergson's vitalism seeks to explain the transition from a body comprised of nothing but particles, to the phenomena which appear in the living body, which is much more complex than any one particle. Bergson's vitalist explanation was an attempt to understand emergence using an additional force. The vitalist view was rejected and scorned as the years passed, and so Bergson lost the credit as the person who identified the importance of emergent processes when observing complex systems, and perhaps even in understanding time itself.

Bergson, following his vitalist view, developed the view which sees a dynamic component in time. He perceives time intuitively and attempts to find the definition of time from one's internal experience rather than the rationalist and scientific approach. He does this while distinguishing between the concept of time and the concept of space. Space, which according to him should be examined as a physical entity, through a mathematical lens, can be divided into an infinite number of discrete components, however this isn't true of time. Time has a "duration" (Durée), it is a renewing, active and creative flow which cannot be divided into discrete moments. The intuitive "duration" does not match "spatial time" which physics describes, including in relativity theory, as an entity which can be divided into frozen, separate moments. The "duration" expresses constant movement and dynamic flow which can only be understood through the intuition. While this view may match our intuition, it's hard to see how it can be applied in order to draw a clear and more "scientific" picture of reality without once again falling into the same paradoxes and discrepancies which have accompanied us until now. In a debate which took place between Bergson and Einstein in April 1922, the two presented their opposing views of the concept of time - Einstein as the representative of physical, static time, in the spirit of Parmenides, and Bergson as the representative of intuitive, dynamic time, flowing along Heraclitus' river. When it comes to concrete results, the relativistic school's physical view became the dominant one in the scientific and philosophical community, but Bergson's spirit has never left us.

Alfred North Whitehead, one of the 20th century's greatest philosophers, will lead us to the next stage in understanding time. Whitehead was a very private man and unfortunately instructed his family to destroy his writings after his death. Whitehead's biographer noted in the start of the biography he wrote on him that "no professional biographer in his right mind would touch him" however Whitehead still had a huge impact on philosophy, including the discussion on the concept of time. He even published an alternative theory to Einstein's relativity theory, and its predictions have proven themselves for a few decades until it was disproved in the 70s of the previous century.

Whitehead raised the idea that "things" (matter, atoms, etc.) aren't the fundamental components of reality, but are rather a result of a comprehensive, fundamental process. The most fundamental unit of reality isn't matter, but "events". Following Bergson's view, Whitehead said that every such fundamental "event" has a duration and is not a static moment. Time and space are, according to Whitehead, a changing and relative framework which is determined by these events. The events are the fundamental components of reality, which can be seen as atomic events, and space and time are a result of the relations between them. Every such fundamental event includes within it its relation to all the other fundamental event units in reality. This fundamental reality is in a constant state of development or emergence, and the relations between the events are also constantly emerging and developing. So according to Whitehead, nature isn't a picture of atoms moving through vacuum but a structure of developing events. Reality is a process.

Anyone who's been following the story up until now probably sees a great deal of similarity between Whitehead's view and that of Leibnitz, and indeed they do have a lot in common. Leibnitz' monad is very similar to Whitehead's event. Whitehead's view isn't easy to stomach, and certainly isn't intuitive, however, later on I will try to connect between insights arising from quantum theory and complexity theory in order to construct a proposal for a clearer and more visual picture of the infrastructure of reality in our world.

An additional insight that can be raised following Whitehead as well as Bergson has to do with the concept of emergence. Whitehead and Bergson included the additional components of "consciousness" and "Élan Vital" into their explanation in order to explain the "life" we see in matter and the emergence of moment in reality out of all possible moments. However, in my opinion, these attempts are actually alternative explanations for the emergence of processes, which can be incorporated into their views and replace the vitalist and consciousness-related explanations. According to the theory I'm raising here, the infrastructure of reality is a collection or network of moments, each of which have a duration or a fundamental process; this duration is what causes the emergence of certain moment - out of all possible moments - in the reality we experience, and can explain some of the mystery in the concept of time which we haven't been able to explain yet. I will develop this idea in the following chapters.

To summarize this chapter, today we understand that nature includes emergent processes. Out of the chaos, the components which make up the system begin an interaction process of communication and synchronization which eventually creates spontaneous order, new phenomena and organization without a directing element. Unlike what we learned, nature isn't in a state of descent in its level or order, nor is it converging to an equilibrium, but rather it is a constant process of creating order. In an exponential process which cannot be accurately predicted, large and small changes create relations of interaction and synchronization which create the emergence of order and the development of a higher level of organization. This produces properties which don't exist in the system's fundamental components.

An ant running around at random and finding food transfers the information to another ant, through chemical processes of synchronization, and she in turn transfers it to more and more ants.

At a certain point the chemical synchronization network moves on to the next step and undergoes a "phase transition", going from a collection of random relations to an organized line of ants making sure to bring the food back to the nest. The organized system that is the line of ants is in a constant state of change, which is why it has a high degree of adaptability to changes in the environment's conditions. There will always be that one ant who, in its random mode of behavior, manages to find the way to bypass the obstacle and create, using an identical process, the bypassing of the entire line.

A "phase transition" is an important step which explains how a random collection of communication between individuals becomes, eventually, a system which develops order and organization in a non-linear way. That is, the development of order isn't a continuous process which progresses consistently and linearly, it is rather a process which begins as a linear transition from stage to stage and through a quick phase transition, becomes an exponential process in which order and organization quickly emerge. This is what we see, for example, in the case of ice forming when water's temperature goes down to zero degrees.

The most common structure in the universe when it comes to complex system is that of a network. The network is a special case of a complex system, however it has additional properties which can serve as the basis for a more developed explanation of processes in nature. In the next chapter we'll get to know the wondrous world of networks in the universe.

CHAPTER 11:

Nothing but Net - Network Theory

Six Degrees of Separation - The Discovery of the Networks in the World

Lately, the great amount of concentration required from me in writing this book has led me to take ever growing breaks and waste many hours on social networks watching cat videos and pictures of dishes served in restaurants, and promoting the business and ideological interests of various people whom I don't know.

Social networks on the internet are a classic and well-known example of the concept of networks. Each one of us is a node in a network connected to additional people (nodes) through various connections. However, the structure created is much more complex than networks we've known about in the past. A fisherman's net, for example, is a collection of nodes where each one is connected to a fixed and known number of nodes adjacent to it, and the interweaving of all these nodes together creates a classic network structure. There are two significant differences between a social network and a fisherman's net. Firstly, the connections in a social network are much more varied: they can be formed between two very distant points, and each node can have a large amount of connections. The second difference is how dynamic the network is. A fisherman's net never changes once it's been weaved - the nodes in the network will always be connected to the same nodes in the net, never changing. The social network, however, is constantly changing; connections (friendships) are created and disappear regularly, and as more people join the network, new nodes are constantly being created: In terms of the visualization, the term "network" can be confusing since most networks we know are nothing like a fisherman's net, they are complex structures of connections of various distances and scales. The image created is more like a tangled bush whose branches are interweaved in countless points in differing levels of connectivity.

The concept of a network, which seems natural and familiar, didn't have any scientific significance in terms of our understanding of the world up until recently. Only in the 21st century did we begin to understand that a "network" is a fundamental concept in our universe. The network understanding of our world, from an economic, sociological, biological and personal perspective, is the revolution of the period we're living in. There are countless examples of networks around us, from the neuron network in our brain to the network of cells in our body, through social networks, economic networks between companies and countries, transportation networks, electricity grids, communication networks, and of course, the internet, and much more.

The developing computing capabilities in the late 20th century allowed us to create simulations of network growth and development, and so the development of virtual networks began being studied, as well as phenomena observed in networks in the biological world. At first researchers focused on a computational study of random networks, that is, networks whose connections are created randomly rather than according to a particular regularity. In such networks, each node had a similar number of connections, similar to a fisherman's net, and the number never exceeded a certain range. However, researchers quickly understood that most networks we find in nature and society don't behave like networks created by a computer - the range of each node's number of connections is usually much larger. The reason for this is that networks in nature apparently operate according to a certain regularity and are not entirely random. So in most networks, any node can have a very different number of connections, which is why they are called "scale-free networks".

Another interesting thing which was discovered has to do with the ability to transmit information in a network. When we take a network similar to a fisherman's net and try to transmit information through it, this could take a while, since each node can only transmit information to adjacent nodes to which it is connected. The information is transferred in a consistent process from node to node until it arrives at its destination. We always knew that as we add connections, the ability to transmit information increases, but we didn't realize by how much. What was eventually discovered is that each connection in a network which connects the node to non-adjacent nodes increases the ability to transmit information in an exponential way, that is, at a much larger scale than the number of connections we added in practice. Each connection added to a node significantly increases that node's connectivity and thus the connectivity of the entire network. The reason is that each connection to an additional node, brings all the nodes already connected to that node closer. For this reason, networks which included connections between distant points were called "small-world networks". If I'm connected to all my acquaintances in Israel and I'm interested in transmitting information to the US, it would naturally be a very long process. However, just a single connection of mine to someone in the US, who is connected to a network of friends there, can make the transmission of information significantly easier, and make our world smaller.

The first researcher who led to the fast breakthrough of network study was Stanley Milgram, a psychologist in the University of Yale, who conducted an interesting social experiment in 1967. He sent 160 identical packages to random people in the US and asked them to give the packages to a certain contact. If they knew the contact personally, they had to hand over the package directly, however if they didn't know the contact they were asked to transfer the package to a person they do know personally, who they think has the highest chance of known the recipient. Milgram's estimation was that it would take hundreds of people to transfer the package all the way to the end, and that most packages would never arrive at their destination.

When the reports were in, Milgram was amazed to discover that 42 packages had arrived at their destination and that the average number of people through which a package had passed was just 5.5. Years later, following a successful play and film called "Six Degrees of Separation", the term became famous - even though Milgram himself never used it - and became a symbol of the fact that we live in a small world, as the play's protagonist says:

"Everybody on this planet is separated by only six other people. Six degrees of separation. Between us and everybody else on this planet. The president of the United States. A gondolier in Venice... It's not just big names. It's anyone. A native in a rain forest. A Tierra del Fuegan. An Eskimo. I am bound to everyone on this planet by a trail of six people. It's a profound thought... How every person is a new door, opening up into other worlds..."

And this was much before anyone imagined the internet or Facebook.

If so, the interaction between a group of individuals creates a network, which includes complex structures capable of transmitting information efficiently, and its process of growth is characterized by regularity. The study of networks today examines the ways in which networks are created and developed, their practical significance in our everyday life and their effect on the nature of the universe. My goal is to show later on that the regularity found in the behavior of networks and the way they develop is the missing link in the understanding of the concept of time. When we understand the different characteristic rules from the behavior of networks and try to apply them to a network which I will propose as a model of the universe, we can understand what time is and how it develops from one moment to another, only in one direction.

The Winner Takes It All (Almost) - The Power Law

We all know the "stars" who walk among us, those people who enter a social event and immediately find themselves surrounded by many people who know them. Any attempt of theirs to cross from one side of the room to the other is accompanied by many stops to greet someone or make small-talks. Today we hardly see such situations since most social activity has moved to the virtual world, and there, the "stars" are those with the highest number of "followers". However, compared to the number of people connected in social networks today, such nodes are rare, and there are very few people who have a very high number of followers. A node in a network which holds a large amount of connections is called a "hub". Every scale-free network has hubs with a significantly larger number of connections compared to the rest of the network. Examples of such hubs are airports in the world's largest cities which are connected by airlines to a large number of other airports; the internet's main search engines, molecules in a living cell which take part in most of the cell's chemical reactions; or tycoons, which constitute significant nodes in the network which controls the world's capital.

As mentioned above, the networks observed in nature and social life presented a non-uniform structure: there are scale-free networks which have nodes with large amounts of connections - such as Google in the internet - but most nodes have a relatively small number of connections. However, this structure was not the end of the story. Albert-László Barabási, one of the first researchers of networks in the nineties, discovered in his research an important property which explains the unique structure we find in networks - Preferential Attachment. This property isn't obvious and doesn't appear in the development of random networks tested in computational models.

Supposedly, it is logical and expected that the number of connections in a network's nodes would be varied, and this has no special significance. This world has many parameters which receive various values within groups, such as the height of human beings or the grades received in the SATs. This variety is called a normal distribution and is described by Carl Friedrich Gauss' famous bell curve: most individuals in a group will be found in the vicinity of the average, and few individuals are found in the areas of very high or very low values. Much like people's height, for example, we find most people in the average height area, and very few people who are very tall or very short. In the early days of the study of networks, such a distribution was expected to be found in this field as well, however when researchers began mapping networks, they noticed that the number of connections doesn't match the normal distribution but rather behaves according to a power law. A power law is a mathematical law, but the mathematical explanation isn't of interest to us; what's important is understanding that a group of individuals which behaves according to a power law is very different from a group of individuals which behaves according to a bell curve. In such a group, most individuals will have very low values and a small number of individuals will have very high values. For example, if people's height was distributed according to a power law, most of us would be 1 meter tall, but from time to time we'd encounter a person hundreds of meters tall.

Most networks are characterized by a power law behavior. For this reason, in the internet, which is a scale-free network, we see very few websites such as Google, which are connected to an astronomical number of connections, and most websites have a very small number of connections. Similarly, there are very few airports which are connected to almost every airport in the world - most airports have a relatively small number of connections; similarly, we see that 99% of the world's capital is held by 1% of the population, and that most individuals in the population have a very small share of the world's capital.

When we look at such hubs in various networks and understand their significance in the control of information in the world (Google) or capital (tycoons), we understand the need to analyze and study why such gargantuan hubs develop in networks. What is the regularity of the development of such hubs? How can their operation and growth be regulated? And how can we prevent them from taking over the entire network?

As part of the attempt to understand the development of such hubs, an interesting process was discovered, one which appears in many networks. The technical term for this process is called "Preferential Attachment" and in the words of ABBA - "The winner takes it all". To be precise, it's better to say that the winner takes most of it, or that "money sticks to money". Preferential Attachment means that the more connections a node has, the chance of it making additional connections increases. The more social connections someone has, the more likely he is to make connections with additional people who want to connect with him; the more control a tycoon has over a bigger amount of capital, the easier it will be for him to take control of additional capital; the more connections a website has to websites in the internet, the bigger the number of additional websites which connect to it. This process explains why the network has exceptional nodes with an extremely large number of connections: the more their number of connections grows, the bigger it becomes, in an exponential and non-linear way.

This discovery has two significant implications. Firstly, we see that there's a certain pattern which gives preference to certain parts of the network over others. The network isn't uniform nor is it democratic; it has areas and individuals with significantly more power and influence compared to others. This difference of certain individuals in a network should arouse the attention of anyone trying to understand why, as part of the interaction of a large collection of individuals, certain unique phenomena are observed rather than other, or in other words - why is order created and why this particular order. Another implication, which will lead us to the next stage in the story of network theory - is the acknowledgment that networks have regularity - something which wasn't obvious up until then. Similar processes appear in all kinds of networks, and so we can start talking about networks in an inclusive way and try to build concepts and a language which will serve as the foundation for scientific research and the application of networks in various fields.

If so, we learned that the more connected a node is, the higher it's chance of getting additional connections, and so, in the process of "the winner takes it all", the hubs become more and more connected and become factors with large importance in the network, both in transmission of information and in the network's stability. The internet is the classic example of the development of a large network of information which began from nodes with just a handful of connections, which went on to create random and intentional connections. At some point, ordered processes of transmission of information began developing, and we witnessed the appearance of nodes with a number of connections that was much larger than the other nodes in the network. The network became so complex and connected that in an emergent way, a collective consciousness grew out of it, which constitutes the largest database of human information in the world. The internet, which indisputably controls our lives today, may include hubs with a huge amount of influence on the transmission of information and the structures which develop within it, but it developed without any planning, directing or overseeing agent. We will now try and understand how all this happens independently from within the network itself, and how new emergent phenomena are created which do not exist in the components which make up the network.

"The Invisible Hand" - The Growth of Networks and the Emergence of New Phenomena

"I think therefore I am" was the most fundamental insight which Descartes arrived at. Thought is the only thing which I cannot doubt. However, what is the source of thinking? Where does our consciousness come from? This question hasn't been answered yet to everyone's satisfaction, and it may never be answered. Does consciousness have a physical origin, or is it a phenomenon of a mental entity which is separate from the body? From the perspective of science, our goal is to discover how an emergent phenomenon of consciousness is created from the neuron network which makes up our brain, a phenomenon which cannot be found directly in the brain's physical components. For this purpose, we must understand how complexity theory and network theory can provide a theoretical template for describing the phenomenon - that is, to understand the regularity of a network's growth and development.

Network operate based on two primary rules: growth and

Preferential Attachment. Each network expands by adding new nodes, and nodes prefer connecting to nodes which already have many connections. The probability of a certain node receiving a new connection is proportional to the number of connections it already has.

Networks in reality aren't static but rather dynamic, in the sense that they are constantly developing and changing. The fact that they are dynamic can stem from an activity which is supposedly random, like an ant running around from side to side without a defined direction or a firefly which turns its light on and off at a certain rhythm. However, if each component in a network goes on carrying out random actions, we will never see any development in the network. There as to be an additional mechanism which leads to the connections between components having value. This mechanism is the transmission of information between components, a feedback loop based on information flowing through the network. In the case of the ant, the information is pheromones that the ants excrete and the feedback is the reactions of other ants to the pheromones they receive. In the case of the firefly, the information is transmitted through the light it spreads and the feedback is the reaction of the other fireflies to the light they receive. In the case of the stock exchange, information is the acts of buying and selling by every component in the market, and the feedback is the reaction of the other investors in the market. In the case of the human brain, information is the messages transmitted through electric pulses and chemical interactions and the reactions of the other neurons and synapses to this information, and so on.

That is, in each network there is a transmission of information from node to node, and the various nodes react to this information. Without a mechanism of this kind, the network would remain static and wouldn't develop and this mechanism is the source of self-organization and the emergent order we encounter in nature's networks. In order to understand the complexity of the network and the way in which order and self-organization emerge, we must examine the independent and random activity of components, the dynamic between them and the interactions between the nodes and the network's connections.

We can do so using a simple model which was developed in the theory of computation called "cellular automaton". A "cellular automaton" is a network of cells in which each cell is a simple unit which can be found in one of two states when reacting to the state of adjacent cells.

In order to understand this model, imagine a chess board in which every square can change its color from black to white according to a certain regularity, which has to do with the state of neighboring cells. For example, we can decide on a rule such as "if the two cells adjacent to the cell are black, the cell becomes black" or "if the adjacent cells are white the cell becomes black", and so on. When we take such a model and let it run according to a system of rules, we get a developing image of cells turning black and white with a certain dynamic. Note that this is precisely the mechanism I discussed before: a collection of components (cells) which operate randomly and only a system of rules which govern their interaction with other cells can cause them to react in a certain way rather than another. Their change also affects the cells connected to them, according to the same set of rules. We begin with a certain configuration of the "chess board", that is, an input of a collection of cells colored black or white according to a certain "template", and in accordance with the network's feedback rules we receive a different configuration every moment, until eventually we receive a final configuration, which is the output.

This model highlights another important property of networks, which is the fact that the relationship between the network and its components is mutual, that is, the components influence the network and shape it, and the network itself influences the components and shapes them in return. The interaction is between the components themselves and between the entire network and the components themselves.

One of the leading researchers of cellular automata is Steven Wolfram, an eccentric mathematician and physicist who published these models to the scientific community and the general public. Wolfram, who published scientific articles as early as 15 years old, and wrote a PhD when he was 20, published a book called "New Kind of Science" in 2002, which includes 1,200 pages and which was written over the span of a decade. In this book, Wolfram analyzes models of cellular automata as part of the attempt to propose a new science which would provide explanations for the universe's most fundamental questions. His work remains controversial to this day, however his contribution to the adoption of network language and concepts from complexity theory by various fields in the study of the universe and society cannot be ignored.

A certain kind of cellular automaton is called a random Boolean network. A Boolean network includes nodes and connections between those nodes, and in each stage, every node can be in one of two states. However, unlike the cellular automaton which I described above, in a Boolean network, connections don't occur only between two adjacent cells, but between distant cells as well, and each cell can have its own rules of operation.

The person who studied such networks, mostly in the biological world, in the context of networks of living cells, was Stuart Kauffman, another important figure in the study of networks and complex systems. Kauffman was one of the first researchers in the institute of complex systems in Santa Fe, which remains that most famous and esteemed center for the study of complex systems to this day. Kauffman mostly studied the development of spontaneous order in systems and networks, and his 1993 book "The Origins of

Order: Self Organization and Selection in Evolution" is a milestone in the understanding of complex systems. He found that as a system becomes more complex, and more connections are made between its nodes, so will a spontaneous order begin to develop throughout the entire network. The system goes through several phases in its development. When a system begins developing and forming connections it initially displays a fixed or periodic behavior. The system is sensitive to initial conditions, which is why later on, as it creates more connections, it displays a random behavior similar to chaotic, unpredictable operation. That is, every dynamic network which creates more and more connections will eventually arrive at a seemingly-chaotic behavior. However, Kauffman's truly interesting discovery is that a living organism which wants to keep on developing and living cannot be in a frozen state in the system and cannot be in a state of chaos either, since in the former state it would become atrophied and die, and in the second state it would devolve into lifeless chaos. Kauffman found an intermediate stage between fixation and chaos, which he called "life on the edge of chaos". The expression "on the edge of chaos" was coined by Christopher Langton, a researcher in the field of computer science, who came to know this phenomenon in his world. When we begin connecting nodes, they slowly form connected pairs or groups, hubs appear, and eventually a single additional connection causes a "phase transition". All of a sudden, the entire system becomes connected at a much higher level, and a new order which didn't previously exist is created. Any random network arrives at some stage at a phase transition, and this stage takes place when the system is between the fixed or periodic pattern and total chaos. At this stage there's a little chaos but it allows for the network's development into its next ordered stage through the phase transition. The phase transition will not occur if the network becomes a little more complex, arrives at total chaos and loses its order.

Kauffman, a biologist and expert on evolution, understood that a living body can only stay alive if it can remain at the stage before chaos, on the edge of chaos, but not if it's in a complete equilibrium, since in such a case it atrophies and cannot develop. Only a system which has a certain degree of chaos can undergo a phase transition and create new order and properties which didn't previously exist. As Kauffman writes in his book "Investigations":

"an organism or organization needs to be poised in an ordered regime, near the edge of chaos, where a power law distribution of small and large avalanches of change propagates through the system such that it optimizes the persistent balance between exploration and exploitation on ever-shifting, coevolving fitness landscapes." Kauffman revealed to us that in their development process, networks can arrive at a certain stage which is the cause of the development of order within them. This stage on the edge of chaos includes some order and some chaos. Order is important to prevent the system from collapsing to total disorder, and chaos is important as an engine for the important dynamic in the connections of the network's components, which will eventually lead to a phase transition and the development of order and new emergent phenomena which do not exist in the system's components.

We therefore learn that networks are made up of components which are nodes of dynamic connections, whose number is ever-growing as the network grows and develops. The regularity of transmission of information in the network and the feedback which affects each of the components are what cause the development and the dynamic nature of the network, and eventually lead to the transition between the various stages: from an initial ordered state which lacks an equilibrium up to a state of total chaos. In an interim state, between equilibrium and chaos, the system is still relatively ordered but already includes some chaotic components. At this stage, systems undergo a quick phase transition and create a new spontaneous order which hadn't existed before, and this causes the emergence of new phenomena and properties which do not exist in the components which make up the network. That is, networks which reach a stage on the edge of chaos create an emergent order. If we look again at a swarm of ants, as a classic example of a complex system or network, in the first stage we find that the ants are moving randomly. In their movement they excrete pheromones, which are the connections between the ants, who are the nodes in this network. Each ant responds to pheromones she receives from the other ants in the network. The bigger the number of ants involved in this interaction, the bigger the number of connections between the ants in the network. At a certain stage the level of connections creates a new order in the ant swarm's network. Order can be expressed in the construction of underground structures which house the ants and their offspring and in which food is stored, or in the creation of a complex supply route which leads to distant locations in which food was found. As we mentioned above, according to Kauffman, life itself is an emergent phenomenon which appears out of the complex network of cells in the body. Consciousness, too, appears as an emergent phenomenon, where the neuron network crosses a certain threshold of connectivity in the network, and so on.

Our stop in the world of complexity and networks in the quest for the concept of time is not coincidental. Later on, I will try and present an innovative and highly speculative view, which claims that the universe itself is built as a network of configurations and acts as a complex system according to the regularity of scale-free networks. In fact, I will attempt to raise a conjecture that the universe consists of a network of moments: each such moment is a configuration of the universe and constitutes a node in the network, which is undergoing a process of ever-growing connectivity. When a certain node undergoes a "phase transition", each such moment (and all of time with it) is realized in reality. I will therefore raise the conjecture that time is an emergent phenomenon which appears in this network of "static" moments.

Feeling confused? No matter, a certain level of chaos is a requisite when it comes to creating new order and moving on to the next stage of development.

Life On The Edge Of Chaos - Insights On Life Itself

Before we continue in our quest for time, I want to pause for a moment and try and connect some of the insights we've raised in the previous chapters to life itself. How can we apply these insights in processes we run into in our personal lives, and in the business framework in which we operate? How many times in your life have you encountered theories, practices, life advice or spiritual doctrines who offered an ultimate answer for the processes we see in the world (personal, economic, sociological, psychological, etc.), or have offered a way to deal with the world in order to achieve a certain goal? How many times have you tried applying these approaches only to find out that something's missing? Eventually you were left with the gap between the elegant and promising phrasing and the result in practice. There's always something missing. We remain without full answers, with the feeling that the world is a lot more complicated than the way it's presented in academia, religion, or some spiritual doctrine. We find ourselves surprised, and insufficiently happy.

The western view, which was based on logic and reductionism, has influenced the human world view by providing the following guideline: if we understand the problem's components and the laws according to which the world operates, we can understand any system (biological, economic, social, personal) and predict how it will operate in the future. According to this view, we can always construct a theory which identifies the components which make up the system, the initial conditions and the rules according to which it operates in order to develop and progress to the next stage. Such a theory can allow us to predict how any system will behave and how we should prepare and plan our life in order to face future developments and achieve our goals. This is the foundation of the western view of understanding the world. This approach has led to the scientific and technological flourishing of the western world and promoted the development of human knowledge to the impressive level we find it in the 21st century. This approach has also seeped through into all walks of life, including our personal and economic world view. However, this approach, and the theories whose development was based on it, only allow us to see the world in an approximate and average way. We can predict how the world will behave, but not 100% accurately. This kind of knowledge is good enough to get us to the wonderful places we've gotten to, but there's something missing; something that will clarify what really happens in complex systems and how they develop, and why we always find ourselves surprised and feeling like we missed out.

Three significant processes which took place in the 20th century have completely changed our view of reality:

Firstly, the understanding that the world is random and has an inherent component of uncertainty, which doesn't stem from a lack of information. Quantum theory has taught us that the world at its very base includes components which don't have certain properties in advance, and that they only appear at random for no reason.

Secondly, the understanding that a system which includes components affecting each other is highly sensitive to initial conditions. Chaos theory has taught us that we can never accurately define initial conditions, and so any observation of reality is approximate and will never allow us to accurately predict how the system will develop moving forward.

Thirdly, the understanding that nature includes systems which can process internal and external information and develop into very high levels of order, without the need of a planning or overseeing agent. Such systems are based on three principles: random or semi-random movement of components, transmission of information between components (communication, synchronization, physical contact) and feedback between the components and the system itself. The combination of these three principles causes the development of emergent order and organization in the entire system.

Based on the understanding that the world isn't entirely deterministic, that is, that there are random processes which don't always have an identifiable cause, and based on the understanding that complex systems (and almost anything we encounter is a complex system) have certain properties and rules, and that they're different from anything we've known so far, the implication is that in order to better understand the world, we must understand the properties and regularity of the combination between the random and non-random behavior of the system's components, and the way order emerges from this behavior, throughout the entire system. If we can understand this, it will do wonders to our ability to understand the world and navigate through it.

If so, what have we learned in the past few years and how can we use these insights in real life?

The world around us, as well as social, personal and economic processes, has some important properties we should know and take into account:

1. **Small changes can lead to huge effects.** Each action we carry out, whether big or small, can have a huge effect, although we can't accurately predict what the result will be, and how big it will be. This understanding gives us a lot of power which allows us to influence the world, but this power

isn't really under our control. It's important to plan and act, but one must remember that there's never a single correct way, and that we will always have to respond to and face the challenges we encounter along the way.

2. **Order and organization are emergent without a planning or directing agent.** Even if we plan and try to manage the process as it develops, there will always be order and regularity created from the interaction between the system's components over which we don't have full control. We should know how to react to the emerging order rather than fight it. We must remember that the influence is mutual: the components form the collective but the collective also defines itself and its limits and influences the components in a circular and bi-directional feedback loop process.

3. **Any plan requires flexibility.** Since we can't make the perfect plan, we should always maintain flexible margins of randomness. When the next crisis comes (and we can never fully prevent it), we'll be able to face it, not because of the original plan, which will collapse in light of the crisis, but thanks to the creativity, irregularity and flexibility we will leave in the plan's margins. A plan that's too rigid will break against any obstacle, but a system with disordered margins will be able to adapt itself to the next challenging situation.

4. **The norm is chaos and crises, not peace and equilibrium.** Due to evolutionary needs, humanity tends to ascribe an order and stability to the world which don't always characterize reality itself. An animal who's unsure of his next step will not move and therefore not survive. However, the ground is always moving, often slightly, sometimes more (in

which case it's called an earthquake). Man tends to ascribe his crises to the past, expecting his future to consist of a peaceful equilibrium. However, reality slaps us in the face every time. The world at its foundation isn't stable and is in a constant state of change and circular transitions between times of crises and times of peace. Spiritual doctrines tell us we should search of peace and a disconnection from the noise of the world around us to achieve happiness, but people going down this path only experience momentary happiness. Only the understanding that chaos is the essence of the world and developing the ability to handle it and live through the chaos can bring about a true, lasting peace.

5. **Development comes on the edge of chaos, and not in states of equilibrium.** The term "on the edge of chaos", in any complex system, is a central factor in the understanding of a system's development to higher levels of complexity. Huge corporations have vanished because they didn't know how to develop to the next stage, and so withered away. Systems which are too ordered and in a high state of equilibrium are atrophied systems since they lack the energy needed to develop and create a new order which can handle the next crisis. Systems which are too chaotic, on the other hand, collapse and disappear. The only way to develop is to be on the system's edge of chaos, in the area in which there is no perfect peace and equilibrium, in the transition between total order and total chaos. Systems can develop and adapt to a higher level of organization and survivability only when there's a certain degree of chaos which creates the new interaction which will lead the system to the next stage. When on the edge of chaos, the system goes through a phase transition, often very quickly, which causes the emergence of something

new. Something which didn't exist before, not in the system as a whole nor in any of its components.

6. **The winner takes it all.** Complex systems mostly function as a network of interactions. One of the primary properties of networks is the development of nodes with higher levels of connectivity compared to other nodes (scale-free networks). In each such network there is a "power law", that is, there will always be a small number of nodes with a huge amount of connections and many nodes with small amounts of connections. This inequality just keeps on growing, since the more connections a node has, the more connections it can form. The development of this structure in complex systems cannot be prevented; we should always look at a system and identify these focal points of power. Mapping them in every system is essential since they are the central junctions for transmission of information, they have the largest amount of influence over the other components in the system and they are the focal points which maintain the network's stability.

We were always taught to seek happiness and success in order, stability and quiet; to disengage from the noises of the world in order to find happiness and plan things better so we can avoid crises. Chaos is bad, order is good. Eventually, one of the most important messages we need to apply is the opposite: not to seek out peace and equilibrium, expecting them to last. If we apply this world view, success and happiness will always be momentary, since the world doesn't strive towards equilibrium. We need to understand that chaos and crisis are the norm, and if we can face them regularly rather than fight them, and use them to develop onto the next stage, we can arrive at success and happiness.

CHAPTER 12:

The End of Time and its Rebirth? - Julian Barbour and Lee Smolin

Where do we go from here? The journey's final stage

A 2,500-year long journey is coming to an end. The end of time. And perhaps its rebirth.

The various stops we've passed through in our journey have strengthened our understanding that physics is doing its best to remove the dynamic and "living" element from time and to turn it into a cold, compressed block of static events. Time and time again, time has taken a hit and lost a bit of its power, and today its death throes are noticeable to many people in the philosophical and scientific community. A significant part of the community and its experts have already lost hope and refrained from their attempts at resuscitation, and some have even officially announced its death.

To be honest, most physicists don't deal with the question of whether time exists or not, and there are philosophers who claim that the very question is rooted in a mistake. The physicist deals with research in his lab and views time as another vessel or parameter in his equations, at best. At worst, he ignores time entirely. Physical time is mathematical. It is a coordinate to be ascribed to the observation we're making, just like any other spatial coordinate, along with the mathematical nuances which are unique to time. Treating time as a mathematical entity makes the researcher believe that he's treating time as a parameter with meaning in reality, but this is an illusion - this isn't the nature of mathematical treatment. Granting meaning of existence to mathematical parameters is inherent to scientific work, and at times reality is described using mathematical entities, even when they have nothing to do with reality. Mathematics has developed a lot over the years and has distanced itself from the type of mathematics which has a clear expression in observed reality. At times there is controversy regarding the question whether mathematical models and objects common in scientific work today actually represent elements of reality. I will not get into this complex question, and in any case most of the insights I presented in previous chapters are ultimately based on the assumption that mathematics represents reality itself. The bottom line is that time, despite the evidence which has piled up over the years regarding its non-existence in reality, is still a fundamental tool in scientific work and is at the backdrop of every experiment and observation.

However, our interest in this journey isn't an examination of the practical use of the concept of time in scientific work, but the attempt to understand time itself and the meaning of its existence in the world, as a fundamental entity with a significant influence on the way we experience the universe as a whole and the world around us.

The 20th century has ended and we have three main pieces of evidence for the non-existence of time, which we've already discussed: the **McTaggart argument** has shown that logically speaking, there is no meaning in the continuum of time moving from the past to the present to the future, and any attempt to identify movement from an event to an event in a sequence of events leads to a logical contradiction; the special theory of relativity's **block universe** has proposed a physical argument which disproved the simultaneity of events in the universe, and so ruled out the difference between the existence of an event which is considered to have taken place in the future and one which has taken place in the past. The identical existence of every event in the history of the existence has once again disproved the development of time in each and every moment towards the future; and finally, the **Wheeler-DeWitt equation** has shown that in the attempt to unify quantum theory and relativity theory in order to construct a wave equation for the entire universe gives us a frozen universe, without any progression in time.

This is all well and good, but the human spirit hasn't changed in the past 2,500 years, and the conclusion which gradually became clearer regarding the illusion of time is at odds with our strongest and most basic intuition: time moves from moment to moment and we operate freely in order to change the future and influence it. This unavoidable contradiction between the scientific and philosophical conclusion and our intuitive feeling doesn't allow us to accept the conclusion and go on with our lives as usual; we have no choice but to look for a world view which can explain both the grim conclusions arrived at by science and philosophy, and the strong intuition which stems from our everyday experience of the universe. We may be wrong, and I've already claimed that the universe doesn't owe our intuition a thing, which has often had to break and retreat from its stance in light of the scientific revelations mankind has arrived at. The world is not flat. The sun doesn't revolve around the Earth. Giraffes did not stretch their necks in order to reach tree-tops. God did not create the world in 6 days 5,000 years ago. Time does not exist. Or does it?

The rest of this book will focus on breaking a new path of thought regarding the concept of time, in an attempt to propose a solution - or at least the start of a solution - to one of the most ancient and complex riddles which still remain unsolved. In the following chapters I will construct the foundations of this additional solution based on the knowledge we've amassed so far in physics and philosophy, which has a central role in the current prevalent paradigm, and which is unanimously considered to be a significant component in any future theory. The theory I will lay out in the following pages will also be based on interesting insights published in the 21st century, some of which are rather new. Ultimately, I will combine the three theories which have accompanied us in this book: relativity theory, quantum theory and network and complexity theory, and I will construct a new view, based on the combination of their principles, which breaks the current pattern of thought and can re-open the desired path.

The world view I'll be describing here does not attempt to provide an answer to the question of whether time exists or not; the starting point I begin at is the question of which of the currently existing theories is the most convincing and founded, based on the scientific and philosophical knowledge at our disposal, and what are the most up-to-date discoveries in scientific research which can hint at a direction which will allow us to construct a more complete solution. I take these foundations and insert them into a new pattern of thought; as with any theory that provides an answer to earlier riddles and thus creates a new order in the level of human knowledge of the world, so do I think this pattern of thought will cause the entire theory to create a phase transition which will lead the entire explanation to a new, higher level of order.

If so, we've arrived at the stage in which we must get to know the most interesting and consolidated views when it comes to the concept of time, which were published in the 21st century by two of the world's most esteemed theoretical physicists: Julian Barbour and Lee Smolin. Their views may be somewhat radical, but they're not completely disconnected from the current paradigm, and what's truly interesting about them is that they completely contradict one another. In other words, in the second decade of the current century, the discussion of the concept of time is still creating opposed world views, each of which is awe-inspiring in how detailed and well-explained they are. Let us get to know the Parmenides and Heraclitus of the 21st century.

"Now" is the Time - Barbour's basic principle

Julian Barbour is, in a sense, the odd duck of the world of physical research, and I think he is just as much the first pioneer of a phenomenon and process which are getting stronger and which will receive significant expression in the coming years.

Barbour is an esteemed researcher who has studied the concept of time for many years. However, at an early stage of his career he decided to leave academia and start a translation business which would fund his continued physical and philosophical research. This personal move is a sign of a process academia is going through, and I think more and more researches will make similar moves in the future. Academia is influenced by two processes which hurt its power and status as a lighthouse of human knowledge and wisdom. Firstly, considerations which aren't purely research-related significantly impact the ability of people in academia to focus on the matters important to them and in which they'd like to deal with. Budgetary considerations which are influenced by the interests of the funding entities and personal considerations of influential people in academia are what determine the path academic research will take. Secondly, information is no longer owned exclusively by academia. The sources of knowledge which have been opened and revealed to the general public, mostly thanks to the internet, provide accessibility to any information required for academic research even to those operating outside of academia. In light of these two processes, academic research which focuses on certain subjects can take place outside of academia - and at times, **only** outside of academia. Barbour, who wished to study the subject of time without being limited by various institutional considerations, chose to do so outside of academia. His book, "The End of Time", published in 1999 is a testament to the ability to conduct productive and innovative research outside of academia, which can receive legitimization both within the academic scientific community and outside of it.

Barbour is the Parmenides of the 21st century. It is not a coincidence that this title of his book declares the end of time. Barbour does away with the idea of the "passage of time" completely - as far as he is concerned, the universe is a static universe made out of a collection of moments, like images on a film reel, and movement and change are not dynamic concepts but rather stem from the sequence of static images. I give his theory a central part in our journey since some of its fundamental components are very strong and have even been verified in discoveries and experiments published in recent years, and so later on I will use them to construct the hypothesis I will propose regarding the concept of time. However, I also think that Barbour stops at a certain point, and that his explanation from that point onwards is unsatisfactory and does not allow us to construct a complete scientific theory which can be tested and disproved. Let us get to know the main points Barbour presents in his book.

The starting point of Barbour's theory is that the world is made out of moments he calls "nows". Each "now" is a configuration of the arrangement of matter in the universe. To put it very simplistically, a "now" is like an image in a film reel which includes everything that's happening in the universe in that moment. For Barbour, the universe isn't made out of Newton's absolute space, where matter is organized in different configurations at different moments in time. Space is not a container which contains all of matter but rather the universe is made out of "things" which are organized in different configurations at different times, and those things are the "nows". That is, the universe is made out of a collection of different arrangements (configurations) of matter which include within them what we call space and time. The analogy Barbour often uses is "triangles". In his lectures, Barbour uses wooden triangles and uses them to demonstrate his view of configurations. Imagine a universe which includes just 3 particles. These particles can be arranged in different configurations. Each such configuration is a triangle where each vertex represents the position of one of the particles.

Each "now" is a triangle which describes a different arrangement, a different configuration of the three particles, and includes the space in which they are arranged as well as the time in which they are arranged. The universe, according to this world view, is a collection of triangles which are piled one on top the other. At each moment in time we experience, one of these triangles receives realization in our reality, and so arrangements of triangles are realized one after the other in our reality. We will immediately discuss the question of why a certain triangle appears at each moment rather than another, but in principle they just appear one after the other. In an analogy to our universe, each "now" is a configuration which is much more complex than a triangle, since the universe has a much larger number of particles and matter, but ultimately, each such moment is a different arrangement of all the matter in the universe. The "change" we experience from moment to moment is the gap between one configuration and the next: if we take two different triangles and place them on top of the other, they will not be perfectly congruent, and so if they are realized in reality one after the other, we will experience a certain arrangement and then another, right after. The experience of the transition between one arrangement and the other is perceived as a change in the matter in the universe, however these are two different triangles which appeared one after the other, and the change we see in the universe is merely an illusion. Just as when images in an animation film appear one after the other, the gap between the arrangements in the images creates an illusion of movement and change.

In a way, Barbour's point of view is similar to the "block universe" view we described earlier, but Barbour's theory is much more advanced and detailed. Firstly, for Barbour, there is a collection of "nows" which includes all possible arrangements of matter in the universe. Each different arrangement of the matter in the universe is a potential "now". All the possibilities of arranging the particles in the universe have the potential of being realized - and this is an infinite number of possibilities - but not all of them are realized in reality, of course. Only a very small part of these "nows" are realized as moments in our reality. That is, what we know as reality, or what the "block universe" includes within it as a collection of events which take place in the universe, is for Barbour nothing but a miniscule part of the collection of "nows" which have potential of being realized. Another difference between this view and that of the "block universe" is that Barbour's moments include within them not just space but information on time, and some of them even include information about the past. That is, "nows" include other "nows" from the past in a complex structure, and they include information about the history of that moment and our memories of those moments. This component of information in a "now", Barbour calls a "time capsule".

The time capsule and the wave equation - the realization of the sequence of moments in reality

Barbour therefore describes a world of static, frozen configurations which are just there, and which have the potential of being realized in reality. Most of them will not become realized in reality and when a certain moment does become realized it will include a great deal of information about its past.

The first problem Barbour needs to deal with is the fact that we experience a distinct arrow of time in a certain direction, and that in each moment we experience, there's a lot of information about moments that came before. Time has a consistency when it comes to the sequence of moments. That is, in each "now" we experience, there's a lot of information about the past and previous "nows", about written history and our memories of the past. Each "now" does not include present components exclusively, but always includes within it a whole story about the history of the universe and about us personally, and this always happens in a certain direction from the past to the future. According to Barbour, this information about the past and previous "nows", which is contained in a certain now, is part of the pattern we find in a "now" called a "time capsule": "each static configuration which seems to include consistent records of processes which took place in the past, according to a certain regularity, can be called a 'time capsule'".

A "time capsule" is basically a fixed, static pattern in which information about movement, change and history is encoded. What we experience as the passage of time is encoded immediately using static means in a "now". The entire past is stamped into these patterns of information, and the more detailed they are, the more real the past seems to us. In order to understand how static information can provide us with information about movement, think of a picture of the patterns a sea wave leaves in the beach sand.

This is a static image which describes movement. From this image we can conclude the movement which took place in it: a beach where the low tide left a pattern of static waves in the sand. When we look at the beach sand, we see a three-dimensional static image which includes information about movement. From this static wave pattern, we can deduce the movement which creates the waves on the beach, and clearly this isn't a random pattern. That is, we have an image which consists of a wave pattern which develops according to a certain rule, but the image is entirely static and does not include the component of time. Time is not required in this description in order to receive a dynamic image of movement and change. This image also includes a description of a direction which can be ascribed to the waves. In other words, each "now" has static information which can be used to deduce change and movement. When we experience the "now", which includes a time capsule of several other "nows", each of which includes complex static information about movement, our brain uses them to deduce the change and movement which are supposedly stamped in the information, and an illusion of the passage of time between one "now" and the other is created. The time capsule includes our memories of the past and records of information about the past. The information about change and movement is encoded in a static image the same way audio or video information is encoded on a DVD.

So, the world consists of "nows", each of which is a collection of images which include our memories and record of the past, and it's all contained in one frozen image. Each "now" includes within it a complex structure of additional "nows", and the fact that we're in one "now" and experience within it several "nows" at the same time is the experience of memories which creates an illusion of a timeline that progresses from the past to the present and the future. We're actually in several "nows" simultaneously, one in the present and some in the past.

The second problem Barbour needs to face is the realization of one "now" in reality rather than another. As we mentioned above, the collection of "nows" includes all the possible arrangements of matter in space, the collection of possible moments, from a singularity as appeared in the big bang, up to a "dead" universe in full equilibrium in which all the particles are scattered uniformly. Not all of these possibilities are realized in reality, and so the question arises - why do only the moments we experience become realized in reality?

In order to answer this question Barbour connects his theory with the principles of quantum mechanics and uses the universe's wave equation (the Wheeler-DeWitt equation) in order to explain why one moment is realized rather than another. As you recall, this equation describes the different possible configurations of the universe and the probability of each such configuration being realized. If we connect this to Barbour's theory, the wave equation describes the probability of every such "now" becoming realized in reality. When a certain moment is realized, all other possibilities (other "nows") collapse into a single possibility. According to quantum mechanics, all configurations are possible but the wave equation's probabilities mean that we only experience some of them. So, the question arises - why is it just the moments which give us an experience of continuity and coherence that get realized? How is it possible that all moments were pre-determined, and yet we experience a clear and consistent process of development? This problem can be solved if we understand why the events which create an illusion of consistent development in the universe have a high probability of being realized.

Barbour claims, as we mentioned above, that some configurations include time capsules, that is, information about the "past". The wave function determines the probability of the realization of each configuration, and configurations which include a time capsule have the highest probability of becoming realized. This is the reason they're ultimately "chosen". Time capsules will give us the illusion that "nows" have a history, but this is an illusion as we mentioned above. In other words, "nows" which include time capsules have the highest probability of realization in the framework of the wave equation, and so we only experience moments which include a history and a sense of continuity of consistent and coherent events. This is, in my view, the weakest point in Barbour's theory. Theoretical and unexplained structures, such as Barbour's time capsule, are often revealed to be "ad hoc" solutions, and ultimately give way to a more consistent and well-founded explanation. It's not really clear what a time capsule is, where it came from and what is it about time capsules that causes wave equation probabilities to focus on them. The problem is that this is a key point in Barbour's overall theory, which is why it hurts the credibility of the view as a whole and makes things easier for the theory's detractors. I will describe this problem and other problems in Barbour's theory later on in the chapter.

One of the most beautiful things about Barbour's theory is the combination of the block universe and the Wheeler-DeWitt equation, which describe a timeless world, with the time capsule and various probabilities of moments becoming realized in reality, according to the wave equation. This combination gives us a world devoid of physical time on one hand, and a universe in which time is emergent from matter and space, on the other. The wave equation only describes the development of probabilities, but they have no expression in reality; the wave equation's expression in reality is each configuration's probability of becoming realized, and so, from a timeless equation which describes static configurations, we obtain the development of the realization of moments which include within them the dynamic nature we experience in the world.

Just as consciousness is emergent and emerges from the neurons in the brain, and we can't find it in the neurons themselves or insert it in another fundamental biological theory, the same goes for time. It emerges from a macroscopic collection of microscopic particles, and it cannot be found in any microscopic physical theory. Barbour tried to breathe life into the static view of the universe; he did so with a certain degree of success but in my opinion, he left the work incomplete. Barbour's theory has strong foundations which can serve for the construction of a new, well-founded theory, but it has its fair share of holes and structures which lack a satisfactory explanation. As a physical theory it isn't well-founded enough in order to compete with others as a contender to lead the revolution and replace the current regime. I will now try to indicate the points in Barbour's theory that require reinforcing or an alternative solution which is more well-founded and convincing.

The foundations missing for a revolution - The weak points in Barbour's theory

Even if we accept the premise in Barbour's theory regarding the non-existence of time, there are still some unanswered questions in the story Barbour tells us about the universe and time in the universe. We will attempt to touch on some of the problems and faults we will have to solve in order to promote the theory to a more complete and consolidated stage. I think that one of the reasons that Barbour gets "stuck" at some point in his explanation, rather than taking the next step is, in my opinion, the same reason most physicists haven't made a significant connection between physics and complexity theory. As you recall, complexity theory focused mostly on the world of biology, since it studies complex systems in which order develops. The deep-rooted view in physics is still heavily influenced by the second law of thermodynamics, and it implies that systems only develop towards higher entropy and equilibrium, rather than towards a decrease in entropy and a cyclicality of chaos and order. This view is so deep-seated that most physicists to this day haven't really tried to apply views from network and complexity theory in the physical world, but rather only in its margins. Furthermore, in the academic world, it is still customary to think that such views are not methodological enough and that their research value is low, and so even researchers who're interested in leading such a move have trouble receiving funds and investing their time in it. The academic physical world is only beginning to understand the implications of complexity theory and network theory and how practical they are in formulating theories and hypotheses about the universe - even though nothing is more complex than the universe itself. However, it isn't surprising that as we mentioned above, the view that spontaneous order does not develop in physical systems is built-in to the physical world, and that only in recent years have people began to doubt this.

Before we try to decisively do away with this view, at least in the context of the concept of time in the physical universe, I will go over some of the problems which follow from Barbour's theory.

Firstly, Barbour stopped at a point in which there is still a need to explain the continuity of time we experience and the coherence of the past with the present, that is, the feeling that each moment is a direct and consistent continuation of the all previous moments. In order to solve this problem, Barbour constructs a new physical structure which, in my humble opinion, isn't well-founded enough and which doesn't have real theoretical physical value - the "time capsule". It's hard to understand where this "time capsule" comes from and how it develops. It is a physical component of information which is part of this frozen image, and Barbour seems to take it for granted. Furthermore, the "time capsule" isn't related to any other moment and so it is possible that each moment we experience has an entirely different past, and that our sense of coherence changes between one moment and another regarding various different histories without us being aware of this. This is a world image which raises deep questions regarding mankind's place in the universe, in which there is no one history, no meaning to the universe and our lives and no physical connection between one moment and another. This lack of connection between moments creates discomfort when it comes to our experience of the continuity of time, since according to Barbour's theory, everything we experience as "the past" is an illusion. It is possible that in this moment a particular moment will get realized which includes a past we've never experienced, but its time capsule contains fictional information about such a past, and so our sense of coherence regarding the sequence of events in merely an illusion. The fact that our past is an illusion is not a satisfactory argument against Barbour's claim, but the fact that there is no type of physical connection between the various moments, and that we just "leap" from one moment to another without any joint information is a fault which, in my opinion, increases the gap between theory and our experience in a controversial way, and requires us to re-examine this issue and find a coherence in nature which has a continuity and which includes the transmission of true information from one moment to another.

The additional question which Barbour leaves unanswered is why moments with time capsule have the highest probability of being realized according to the universe's wave equation. What, physically speaking, causes the "now" with a detailed time capsule to have a higher probability of being realized compared to other "nows"?

In other words, I think that Barbour takes the static view of the universe too far and describes an image completely disconnected from the reality of the dynamic world which develops from one moment to another. Barbour describes a world of still images in a dusty pile. At any moment, some law of nature which we don't truly understand, chooses an image from this pile which includes all of its memories and turns it into reality. At each additional moment, all of reality and its history can change. I believe that Barbour does not provide an explanation for the world's dynamic nature and the fact that the time we experience has a true and fundamental coherence which is based on the transmission of information from one moment to another, starting from the big bang up to the end of the universe. The fundamental gap in Barbour's theory is a result of a lack of connectedness between all these moments. Barbour didn't take the extra step in order to arrive at the insight that the universe is a network: a network of inter-linked moments which create a new order that develops spontaneously at each and every moment.

Another fault in Barbour's theory is the description of the way in which the illusion of the passage of time appears in our consciousness. Barbour describes a process which takes place on the axis of time in our consciousness, and thus separates, physically speaking, between the human mind and the rest of the universe. It's impossible for the entire universe to consist of static moments, with the human brain being the only place in which a temporal process which connects several "nows" and creates a continuity between them takes place. If these moments are static, then the human brain itself is also made out of static moments, and no dynamic process takes place within it on the axis of time. The human brain has to be a part of one, singular view together with the universe as a whole, and a complete theory needs to explain why we experience the passage of time without separating between the fundamental principles according to which the universe operates, including the human brain.

Despite the faults in Barbour's theory, and despite our difficulty in accepting this view of the universe intuitively, his view has some strong and well-founded components, and primarily it matches the insights which physics provides us regarding reality. In order to balance out Barbour's radical view, we will present a somewhat opposite view, which venerates the existence of time as a separate entity, which precedes the universe itself.

The rebirth of time - Lee Smolin's theory

On the other end of the scientific community, we find the Heraclitus of the 21st century. Lee Smolin is an esteemed and well-known physics professor, a researcher and one of the founders of the Canadian Perimeter Institute for Theoretical Physics and the author of several books on various subjects in physics. His last two books were inspired by a Brazilian philosopher called Roberto Mangabeira Unger, with whom he co-authored the book. Smolin and Unger have been studying the existence of time together for the past few years. In 2015 they published a book on the subject titled "The Singular Universe and the Reality of Time", and a year before that Smolin published his book "Time Reborn", which summarizes their central claims.

Unger and Smolin developed a theory which combines philosophical and physical elements, which states that time has a primal and fundamental existence in the universe, which pre-dates everything else, including the laws of nature themselves. When I'll be describing the main points below, I will refer to Smolin for purposes of convenience, but the stance being described belongs very much to the both of them.

As you've already understood, Smolin's approach is unique and innovative in the physical world. It is controversial since presently, the physical community holds several very well-founded theories which provide it with very practical tools, and these theories imply a static, timeless world. However, Smolin and Unger's work is based on deep and well-founded physical and philosophical arguments and so should not be ignored - especially since it aims to rescue that intuitive time whose existence we desire so much on a personal and human level.

Smolin doubts one of physics' most fundamental principles, which he believes is one of the main causes of the mistake which has accompanied us since the days of Descartes and Newton. According to this principle, which Smolin calls the "Newtonian paradigm", when we examine a physical system, all we need to know are the system's initial conditions and the laws according to which it operates. If this arrangement and its laws are known to us, we can predict how the system will develop, as well as any future state of the system. This is precisely the mechanistic view we described in the first chapters, and we've already shown that this view has lost its methodological strength following the insights of the 20th century, most of which stem from quantum theory and chaos theory. However, beyond the fact that most of the scientific community still views the world through the lens of this paradigm, Smolin believes that this approach embodies a problem that goes much deeper than any other problem we've discussed so far. The Newtonian paradigm makes us observe a part of the universe, draw conclusions about it and apply those conclusions on the universe as a whole. That is, we can perform an experiment at a lab, deduce a law of nature from it and apply this law to the universe as a whole, assuming it had always existed. How do we know that a certain regularity we've discovered in a lab is relevant and applicable to any place in the universe, and to the universe as a whole? According to Smolin, this approach, as natural and deep-seated in our world view as it may be, is wrong.

Smolin claims that the basic assumption that the laws of nature are identical anywhere in the universe and that any law of nature can be applied to the universe as a whole is a mistaken assumption which leads to a fundamental fault in our ability to construct practical and meaningful physical theories, and it is one of the causes for the fact that we haven't been able to solve some of modern science's biggest riddles and paradoxes for the past 100 years. According to him, the laws of nature can be different in different parts of the universe, and the laws of nature which apply to parts of the universe right now will not necessarily remain that way forever. Smolin calls this mistaken process "doing physics in a box". "Physics in a box" is exactly the methodology, widespread in the scientific community, which carries out research on an isolated system, deduces a regularity from it and applies this regularity to the universe as a whole, for all eternity.

Another problem with the Newtonian paradigm is the assumption that the system is isolated from its environment, and that the results obtained from the experiment are not influenced by external factors. And yet, in an experiment, we can't completely isolate the system, and any result and tested piece of information will always include approximations, estimations and noise, which we identify and separate from the relevant information. But we can always claim that the noise is the true information and that the separation is artificial, and so the results are mistaken and the interpretation of the findings and their adaptation to the theory are mistaken as well. Furthermore, when we "do physics in a box" and carry out an experiment in a lab, we, the observers, are outside of the tested system; but when we apply this system's regularity to the universe as a whole, we insert the observers into the system being described. This method collapses since in order to apply the experiment's conclusions to the universe as a whole, we must observe the universe as a whole as an external observer, which is impossible. The same goes for time - the external observer who examines the system in a lab holds the clock according to which the system is being measured, and the clock is external to the system as well. When we attempt to apply a regularity to the universe as a whole in accordance with processes we've seen in a lab, we measure the system using a clock which is internal to the system, since just as we cannot place an observer who is external to the universe, we cannot place a clock which is external to the universe. It may very well be that this fault, of a watch being placed outside of a system before being inserted into it, is one of the reasons that in physics, which is the result of this methodology, a timeless universe is obtained.

Another assumption made by Smolin and Unger, who they believe is critical for the understanding of physical reality, is that there is only one universe. According to Smolin, multiverse theories stem from precisely this fault in the Newtonian paradigm. When we conduct an experiment in a lab and attempt to understand an isolated system, we can reboot this system, create different initial conditions each time and examine the system's development again and again, in order to define a law of nature based on the results obtained from all of the system's various arrangements. However, we only have one universe with unique, one-time initial conditions, and we can't examine the development of the universe as a whole with differing initial conditions. For this reason, examining isolated systems within the universe and applying laws of nature discovered therein to the universe as a whole is a methodological fault which creates unsolved physical questions. The solution physicists have found to some of these problems is to simply assume that there are many additional universes, each with different initial conditions, but this is a theoretical construct which stems solely from the fault in the Newtonian paradigm. For this reason, Smolin rejects theories which try to solve cosmological and physical issues using many universes.

Smolin's next significant assumption is that time has a fundamental ontological existence in the world, which precedes anything else in the universe. This is a very important assumption since according to him, if nothing precedes the existence of time, then this includes the laws of nature themselves. The significance of this assumption is that the laws of nature themselves develop and change with time: if we assume that time has existence, then the laws of nature must be time-dependent, and we can't assume that time exists while the laws of nature are eternal and unaffected by it. For this reason, we must assume that the laws of nature, the symmetry we see and the constants we assume, are all time-dependent and can change with time. This is a tough claim to stomach for the scientific community, since scientific effort is fully focused on finding eternal laws of nature according to which the universe has always operated, while Smolin claims that the universe has no eternal law of nature.

One claim which reinforces Smolin's theory is the assumption which currently exists in the study of cosmology, according to which the laws of physics we know could not exist in the extreme conditions which were present during the first few moments of the big bang. If the laws of physics can't explain the first moments of the universe, it's likely that at the time they were different from what we know today, or that the laws we know today are incorrect. According to Smolin, the laws of nature change with time, and so the laws we know today are not the same laws which existed in the universe's past and will probably not be laws of nature in the future.

One of the fundamental insights which follow from this assumption has to do with physics' sacred cow: mathematics and logic. Since Pythagoras, mathematics has been treated with reverence, and accorded a primary and even mystical significance. Any argument in modern science since the days of Descartes and Newton was built on mathematical foundations, within the framework of the rules of logic. Now comes Smolin and claims that mathematics and logic don't reflect physical reality and do not precede time itself. A good example for the gap between mathematics and reality in the context of the concept of time is the concept of "movement". The movement of an object in space can be represented mathematically using a graph which describes a change in position on the axis of time, and we get an image of space with a line passing through it, which describes all the points the object has passed through for a certain period of time. This is an image which doesn't include time - we've taken movement in the real world and converted it to a frozen mathematical representation. Now we're looking at moments frozen in time on a graph which indicates a sequence of such frozen moments and claim that this fixed image represents what goes on in reality, that is, that mathematics precedes time itself. Movement's timeless mathematical representation is one of the reasons time was removed from the physical description of the world. When we describe the universe using a model for space which includes a mathematical graph of the change of matter within it, this mathematical model isn't reality itself. The mathematical model does not address time. The model is a-temporal and ever-present, but in reality there's always a certain "now", which is just one point in an infinite sequence of points in the mathematical model, and has no special meaning compared to the rest of the "nows". Is this a mistake which began with Newton and Descartes, which makes us describe the dynamic world using a-temporal mathematical and logical tools to this day? Or is this a fundamental insight about reality itself, which is timeless?

If we take the mathematical model seriously and ascribe a direct relation to reality to it, we lose the concept of time and the difference we feel between "being" and "becoming". There are no longer any emergent processes in reality itself. Reality is just out there, a timeless block of moments. Smolin essentially says that the "block universe", the McTaggart argument and the Wheeler-DeWitt equation all stem from this fault of applying mathematics and logic as representative of reality. However, according to Smolin, the existence of time in reality precedes mathematics and logic, and so these models only provide us with the illusion of a timeless universe.

According to Smolin, only this moment is real. There are no a-temporal laws, and the laws of nature themselves can change with time. The laws of nature, along with mathematics and logic, can only be real at a certain point in time. There is an objective difference between past, present and future. There is no object or mathematical model which is an a-temporal image of the history of the universe, and so Laplace's argument, according to which we can deduce any moment in the past and future based on the state of matter and the laws of causality, is decisively ruled out. It's ruled out not just due to all the reasons we've described, stemming from the uncertainty principle, the subject status of the observer in the universe, and from chaos theory's sensitivity to initial conditions, but also because the universe has no concepts of initial conditions and eternal regularity.

We should be honest, Smolin and Unger make us uncomfortable - and justifiably so.

After we've succeeded in facing the intuitive difficulty of accepting the objective non-existence of time, or at least, after we've understood the arguments of the philosophical and physical theories which support this claim, Smolin and Unger come and rattle the sense we've gradually established using a series of convincingly phrased arguments. Our intuition is once again reinforced and harsh questions arise regarding the insights we've arrived at in the context of the concept of time and the foundations of the scientific methodology and view for the past 300 years.

However, Smolin and Unger's view is, at the moment, merely background noise in the extravagant scientific enterprise created by mankind, and it is hard to say that this enterprise is at any risk as a result of their publications. For now, their view is based on philosophy more than it is based on physics. Smolin, being a physicist, may be trying to sow scientific seeds and proposes several predictions which follow from this view which can be scientifically tested and disproved, however these are still very general statements, which just aren't interesting enough for the time being. The scientific theories of the 20th century are still too stable and well-founded for us to be able to say that we're close to a scientific revolution, and there is a long way to go before we can say that this philosophical view and these elegant arguments are being seriously studied, or that they've significantly influenced the existing paradigm.

When we look at Julian Barbour's world view, contrasted with that of Smolin and Unger, we have to admit the difficult and surprising truth: not much has changed over the past 2,500 years. Can these two world views be reconciled with one another? Can the strong desire to grant time and objective existence be combined with the physical theories which deprive it of such an existence? It's hard to say, for now. As mentioned above, I will attempt to construct initial foundations for such a model. One thing is for certain - in the third decade of the 21st century, we're beginning to feel the ground shaking beneath our feet, and the margins of the physical community are beginning to bubble. What will be the implications of these oscillations on the presiding paradigm? In order to try and answer this question, we will stop for a moment and see how science is progressing and what is the process in which one paradigm is replaced by another.

CHAPTER 13:

R-E-V-O-L-U-T-I-O-N - The Advancement of Scientific Knowledge

Science, Reality, and Truth - Scientific Discovery and the Verification Principle

A well-known epitaph on Sir Isaac Newton by Alexander Pope says: "Nature and nature's laws lay hid in night: God said, "Let Newton be!' and all was light.' This quote reflects the way in which we perceive the scientist as he makes discoveries, and science itself in the last 300 years. The discovering scientist is a heroic figure who sacrifices its life in order to reveal the hidden laws of nature to us, so we can get to know the universe we live in and build practical tools which can make our life more efficient and convenient (and maybe even make 1% of us rich...).

We believe that science's role is to shine a light on the laws of nature and reflect reality. Newton didn't think up any laws of nature, he simply knew how to illuminate and reveal to us that which was always there. What the scientist reveals to us in nature is of course the "truth" about reality itself. Over the past few hundred years, this view has created several cultural heroes known by almost everyone: Galileo Galilei, Isaac Newton, Charles Darwin, Sigmund Freud,

Marie Curie and Albert Einstein. These are examples of icons we all know who discovered new worlds in science, just as the greatest explorers - such as Columbus, Marco Polo and Magellan - revealed unknown territories to us.

But what is a scientific discovery? How does a certain scientific theory appear one day and how does it supersede the previous prevailing theory and take over the realm of science? The more we study these questions in depth, we understand how interesting and complex the process in which science develops from one stage to another is.

By the end of the 18th century, there were two philosophical world views which explained the way in which science reveals reality to us. The rationalists, on one hand, most famously represented by Descartes, claimed that the best way to get to know nature is using our rationality, i.e., through thought. Our thinking comes before experience and so we must first arrive at insights about the world and reveal its truth using the rules of logic. With these insights we must approach nature and study it. The empiricists, on the other hand, claimed the exact opposite: in order to get to know the truth about nature we must approach it in a state which is completely free of thoughts, prejudice and patterns of thought. Our experience of nature comes before rationality. We must perform experiments and observations, and construct models of nature based on their results. What the empiricists and rationalists had in common was the basic assumption that there is one truth and one way to reveal it (of course, each school of thought believed that its way of doing it was the right one).

And then the science of the 19th and 20th centuries came along and began shattering well-founded theories which were perceived as truth up until then. Science was no longer a tool which reveals truth but a tool for developing theories on "truth" which may later turn out to be wrong. Newton didn't light up the universe, he lit up his Principia Mathematica; he presented a theory which turned out to be one theory which may have alternatives in the future. If so, what is "truth"? What counts as science which reveals the truth about the universe, and what counts as charlatanism or theories masquerading as science, which don't reveal any particular truth? These questions will lead us to the understanding of the process of scientific discovery and the progression from one valid theory and another.

In order to save science's status as the discoverer of truth, a clear conceptual framework had to be constructed, one which will clearly characterize and define what counts as a valid scientific discovery which reveals truth about reality and what doesn't. For this reason, following the approach which emphasizes the importance of demarcating what is scientific and what isn't, so we can distinguish between science and metaphysical and theological approaches, the scientific method has itself becomes a characterizing and demarcating factor. That is, if we clearly define scientific methodology, the method, tools and principles according to which we "do science", we can examine each field and determine whether it operates according to this method or not, and then we can define it as science and its discoveries as valid; if it does not operate in accordance with this methodology, then its insights should not be regarded as valid science.

What's relevant for our purposes here is the question of discovery. How is a new theory given the status of scientific discovery, and how does a certain hypothesis get defined as scientifically valid? In the 1930s, the British philosopher Alfred Jules Ayer illustrated the verification principle. According to this principle, a statement can be determined to be scientifically valid only if we can point to a possible experiment (even if only theoretically) which can verify this claim. It follows that according to Ayer a discovery can have scientific validity only if the claim which follows from it can be verified through experiment. Ayer's claim sounds logical and intuitive, but it suffers from a significant fault which we've already known about since the 18th century as the "induction problem", published by the famous philosopher David Hume. Hume claimed that we cannot make a claim based on a specific case and generalize it, that is, we can't formulate a law of nature based on the result of a particular experiment. For example, if we see a white swan, and the next day we see yet another white swan, we can't claim that all swans are white, since one day we may run into a black swan and then our claim will lose its truth value. So, the verification of a specific case, even if it repeats itself and we verify it again and again, cannot be made into a general law of nature since the next attempt to verify it may prove it wrong. It also follows that reproducibility (the ability to conduct a certain experiment again and again while always receiving the same results) is an important condition of scientific research, but it cannot serve as the defining principle of a valid scientific theory. Since we can never conduct an infinite number of experiments in order to verify a claim, the verification principle cannot serve as a basis for determining the truthfulness of a scientific theory.

Let there be light, for now... - Popper's falsification principle

The problem which follows from Ayer's verification principle and the induction problem bothered Karl Popper, a philosopher of science who was active around the same time as Ayer. Popper realized that the claim "all swans are white" cannot be judged as being scientifically valid simply by observing swans who match the claim, since there's always the possibility of finding a swan which isn't white. The verification principle doesn't allow us, according to Popper, to distinguish between scientific and non-scientific theories, and so we need to provide a theoretical definition for a scientific theory (which aims at revealing the truth) which is more precise.

In order to do so, Popper chose an approach which was the opposite of Ayer's and developed the falsification principle. According to this principle, a scientific theory isn't one which can be confirmed, but rather one that can be disproven - that is, a theory can be considered scientifically valid only if we can point to an experiment which could disprove it. Only if such an experiment exists - whether theoretical or one which can be carried out in practice - can we claim that the theory is scientific. The impressive thing about Popper's claim is that it has a strong statement regarding the demarcation principle. It's easy to separate between fields such as religion, for example, whose claims cannot be disproven since there will always be an answer for why a certain phenomenon occurs rather than another, and scientific fields in which we can, relatively easily, define an experiment or phenomenon which can disprove a scientific claim. And indeed, the falsification principle was widely assimilated into the scientific community's views. After Popper it is much easier for us to distinguish between scientific claims and claims which don't meet scientific criteria. Theological claims can no longer be considered to be scientific: there is no experiment or phenomenon which could disprove the claim that every event in our everyday life is a result of divine intervention. For each phenomenon which can be considered to be a deviation from these claims, theology will find an answer which will prevent them from being disproven. Fields which define themselves as scientific have began being tested using the Popperian definition, and disputes regarding that definition have arisen. For example, many claim that Freudian psychology cannot be considered scientific since there are various explanations for mental phenomena, and we can always find explanations for cases which seem to disprove the claims which follow from the theory, within the theory's framework.

Since Popper, science has lost its ability to shine a light of the truth of reality. Scientific theory no longer reflects true reality, it is only true so long as it hasn't been proven otherwise. The theory loses its value of absolute truth and becomes temporary, and if truth is temporary it is no longer the absolute truth of reality as it is. From now on, we should say that science is a tool for creating theories about the phenomena we see and nothing more; while these theories must meet the strict methodological rules, which have been established as scientific, they will always be conditional.

What Ayer and Popper have in common is their understanding of the process of how science progresses from an existing theory to a new one which replaces it. They both identify a significant waypoint which takes us from one theory to another. This waypoint is the existence of a "decisive experiment", that is, an experiment which confirms (according to Ayer) the new theory or disproves (according to Popper) the old theory. As long as a theory hasn't been disproven it is the "true" theory, and as soon as a decisive experiment disproves the prevailing theory, a new replacement must be found.

And yet, the way in which Popper describes the progression of scientific knowledge is lacking and far removed from what happens in practice. Firstly, it is very hard to define "falsification". In the real world, results which don't match the theory are obtained all the time, and there are anomalies which don't match predictions. Is any experiment which yields such results a decisive experiment which can make a certain theory officially disproven? Of course not. Therefore, the ideal world Popper describes does not exist in reality. We will always be able to explain anomalies and experiments which don't match predictions using "ad hoc" hypotheses or by coming up with explanations for why the results aren't valid. Most of the time we won't even know what has truly been disproven - was the whole theory disproven or just an assumption which constitutes a part of it. It often happens that various studies prove something or other, and soon after that, other studies prove the exact opposite. Who can decide if such a study disproves a certain theory, and which part of it exactly it disproves? Secondly, Popper doesn't explain the process through which a new theory is selected as the new prevailing theory in the scientific community. What causes a certain theory to receive the hegemony right after its predecessor was disproven? Popper's description, of science progressing step after step linearly following decisive experiments which disprove previously existing claims, is a lacking and unrealistic description. What's missing in Popper's description is the human and communal element of science; there's something else which takes place in the process of science's progression which makes it much more complex than Popper's naive, linear description.

If so, we've understood that we cannot know which anomalies and contradictory results we meet along the way can truly be considered to be results which disprove an existing theory; we don't know what has truly been disproven as part of the experiment - was it the entire theory or just one of the countless implicit and explicit assumptions which lie at the basis of every experiment? Furthermore, we don't know what causes a certain theory to take the place of the disproven one. So, we shouldn't look for how science progresses in the experiment itself - the answer to this question must be sought in the experiment's environment, in the people carrying it out and in the social processes which take place around the dynamic scientific knowledge in the scientific community.

The first person to include the human component in the process of moving between theories was a French physicist and philosopher of science, Pierre Duhem, who was active in the late 19th century and the early 20th century. Duhem claimed that the "decisive experiment" will never take place in practice, since observing the experiment's results will never be objective, but rather always influenced by background assumptions, and the accepted hypotheses and theories of the time. Any experiment result will be interpreted based on the world of knowledge in which the researcher lives, and so the question when an experiment is a falsifying one is a subjective question. Duhem found the reason for the fact that at a certain stage the scientific community gives up on a previous theory, sets aside all attempts to explain the anomalies which appear in experiments and moves on to the next theory. According to him, this take place based on "good sense" (bon sens). The "good sense" is the community's ability to break out of its own conformity and stop insisting. Duhem claimed that science, which is a dynamic field of activity, is constantly dealing with experiments - some of which confirm the existing theory while others don't meet its predictions. The researcher interprets the various experiments in accordance with the prevailing theory and if it is implied that there is no compatibility with the theory, he finds various explanations for that fact. We've already mentioned two examples for such processes: the "ether" hypothesis and the hypothetical planet called Vulcan. Both were additions made to the prevailing theory which served as explanations for anomalies in cosmological observations which didn't match the theory. There are many additional examples for explanations which turned out to be faults of the prevailing theory at the time, and which disappeared as soon as the theory was replaced by a newer one. The Michelson-Morley experiment, which proved that the "ether" didn't exist, was not a decisive experiment as it is sometimes described in science history books. At the time, various explanations were found for the non-compatibility between the experiment results and the previous theory's predictions, and it was only recognized as the experiment which disproved the concept of the ether 20 years after the experiment was carried out, after Einstein had already published relativity theory. According to Duhem, at a certain point the scientific community uses its "good sense" and realizes that the accumulation of anomalies forces it to give up on the existing theory and look for a new one. The reason that an experiment is never decisive at the time it's being carried out is that the researcher and the community are both heavily invested in the prevailing theory and the assumptions and hypotheses which follow from it, and try to interpret the experiments result within the framework of its language.

While Duhem does not use the term "network", he essentially described scientific interpretation and the progression of scientific knowledge as part of a network of hypotheses, concepts, theories, authority figures in the community and so on. The experiment's interpretation and the decision regarding theories being disproved stem from the network's interaction.

Duhem's view expresses a holistic point of view regarding the scientific process, which operates as a network of knowledge, hypotheses and people. The network's interaction acts on the researcher as he interprets the experiment's results. A scientific theory cannot be disproven in a focused and specific way, it is a process which ends with our "good sense" for decision making. Duhem focused on the researcher's interpretation in the lab, and didn't discuss the community and its role, but he was the first person to include a subjective point of view in the examination of the scientific process. This perceptional component turns science from a body of knowledge which describes reality objectively into a system influenced by trends and the researcher's subjective feelings. At each step of science's progression, decision making is influenced by the scientist's own intuitions, and he is no longer perceived as a machine which linearly progresses from step to step in a manner that's objective and devoid of human considerations. As we mentioned above, Duhem himself did not include the scientific community in his description; this important move in the understanding of the scientific process and the occurrence of revolutions in science will be carried out by Thomas Kuhn and Ludwik Fleck.

The social revolution - Kuhn and Fleck

The most famous science historian of the 20th century, whose name is often mentioned to this day, was the person who introduced the term "scientific revolution" into our everyday language. In 1962, Thomas Kuhn published his book "The Structure of Scientific Revolutions", and popularized the concepts of revolution and scientific community in discussions of scientific development and science in general.

His world view was greatly influenced by Duhem, who we've already met, and Ludwik Fleck who we'll get to know later on and was not entirely original. However, he popularized it and introduced these concepts into academic discourse. One of the most important concepts in Kuhn's view is that of the "paradigm", which describes the scientific world view according to which the scientific community operates at a certain point in time. The paradigm is constructed based on a certain scientific accomplishment around which all scientific activity takes place in an attempt to mimic and complete it. It includes the prevailing theory, the prevailing scientific methodology, the concepts and language they use, the hypotheses which stem from the theory, their interpretations and the central figures in the academic world who support the theory and develop it. An example of a paradigm could be Newton's theory, relativity theory, quantum theory, etc. Following Duhem, any activity, experiment, hypothesis and result is interpreted in the paradigm's framework and through its lens. This activity, within the paradigm's framework, is called "normal science". Kuhn, unlike Duhem, believes that the scientific community, as a whole, is a significant factor in scientific interaction. The community deals with normal science and the paradigm is the template through which the community sees and interprets reality. That is, Kuhn takes Duhem's subjective component and makes it much more dominant in science. Now everything we see in reality is influenced by the entire community, as well as a large array of parameters which aren't objective.

Kuhn agrees with Duhem that Popper's decisive, falsifying experiment is unrealistic. He claims that theories change following a process, which may be clear and revolutionary, but it is also gradual. The revolutionary process begins with the discovery of anomalies and a lack of compatibility between the paradigm and the results in practice. At first the anomalies are interpreted within the paradigm's framework, but eventually they begin accumulating, at which point revolutionary science begins taking place at the community's margins. Most of the community continues adhering to the paradigm and the normal science, but little by little, the revolutionary science which is being conducted by a small group within the community begins receiving reinforcement and grows stronger. When a crisis is created within the paradigm's framework, alternatives to the paradigm are placed at the center of the stage, and the community is already in a stage of transition. At some point, one possibility from the various alternatives begins receiving the entire community's attention, and that's when the revolution takes place, and the community shifts from the old paradigm to the new one. The classic and most famous example of this revolutionary process is the transition from the Newtonian paradigm to Einstein's relativistic paradigm. Newton's theory was the prevailing paradigm for 250 years and all activity within the field of physics was carried out through its lens. Later on, anomalies began to appear, and these were solved using concepts and interpretation constructed from the paradigm, such as the concept of the "ether". Michelson and Morley's results, which didn't match the existing paradigm's predictions, were interpreted by the two as a problem of the measuring instruments' sensitivity, and the physicist Hendrik Lorentz went as far as interpreting the results by claiming that objects themselves shrink as they move through the ether. These interpretations stemmed from a desire to have the results fit the paradigm. At some point, due to additional anomalies, a crisis was created within the community and it was unclear in which direction it was headed. Then several alternatives appeared, from which Einstein's accomplishment broke through, which led to the relativistic revolution. From this moment on, science, as a whole, saw the world through the eyes of relativity theory and the Newtonian view lost its place as the prevailing paradigm.

The main problem in Kuhn's view is the clear boundaries between the various possibilities and his incisive definitions (which may have stemmed from the fact that Kuhn was a physicist and had a tendency of viewing the world that way). Even though he explained the process of scientific development using many subjective elements, Kuhn defined certain terms in a way that was too clear-cut, while in reality the distinctions between them weren't so clear. The difference between correct science and revolutionary science isn't entirely clear, and neither is the distinction between one paradigm and another. According to Kuhn, as soon as the paradigms switched places, they no longer have any common language and one cannot operate within the frameworks of both paradigms at the same time, but in the real world, this division isn't so clear-cut. For example, many scientists today deal with three different paradigms simultaneously (relativity, quantum mechanics and Newtonian physics), which speak very different languages. While Kuhn did clarify many stages of the process and had many important insights, he didn't fully realize how complex this process was, or that it wasn't as unambiguous as he thought. The person who made the next important step was a doctor and biologist called Ludwik Fleck. It is interesting to note that Thomas Kuhn knew of his work and even mentioned him in the margins of his book, and yet Kuhn's physical thinking probably did not allow him to fully take in Fleck's claim, which was a result of the biological point of view in which the boundaries aren't as clear as they are in physics.

Ludwik Fleck's personal story is a fascinating one in its own right. Fleck, a Jew, was born and worked in pre-WW2 Poland. After the Nazi invasion, he was transferred along with his family to a ghetto where he served as a doctor, and even used patients' urine to develop a typhoid vaccine (which was very common in the ghetto). Later on, he was transferred to Auschwitz and Buchenwald and worked in those camps treating syphilis patients and developing the typhoid vaccine. Later on, he would recount that he'd provide the SS soldiers with useless vaccines while sending samples of the active vaccine to laboratories in Paris in which the Nazis reviewed his work. In 1957, after surviving the death camps and serving in different roles in Polish academic institutions, he emigrated to Israel and worked in a center for biological research in Ness Ziona until the day he died.

Fleck's work studying syphilis led him to take an interest in the philosophy of science as well. His world view was described in his book "Genesis and development of a Scientific Fact" published in 1935, years before the sociological study of science was developed, and is now regarded as a sociological view of science which was ahead of its time.

Fleck's new insights regarding scientific discovery and the development of scientific knowledge comes from worlds outside of physics and philosophy. Kuhn was a physicist and historian; while he was influenced by the historical point of view, he also held onto the physical view of the universe as a "machine", and in the clear demarcation between paradigms, and between normal science and revolutionary science. In contrast, Fleck the biologist had a view of science which included organic world views and knowledge of non-uniform, hierarchical structures with no clear boundaries. Thus, for example, in a living cell or a bacteria colony we find non-uniform structures and components with no clear boundaries between them.

When Fleck studied syphilis, he described a concept that doesn't exist in reality. No one has even seen syphilis; what we see are symptoms which we ascribe to the disease, but these symptoms can vary from person to person, and from one point in time to another. When we discuss a disease such as syphilis, we're actually talking about a conceptual structure which is a collection of phenomena which may receive a different name in the future, or may have to be separated into different groups representing different diseases or different conceptual structures. This lack of clarity and fuzzy boundaries between different concepts, which are much more characteristic of biology than physics, allowed Fleck to look at the sciences as a whole from a new point of view.

Scientific fact is a social construct according to Fleck. Thus, for example, syphilis is a human and social concept attached to a certain patient with certain symptoms since that's how the collective defined the concept, and since we see the patient through what Fleck calls "thought style" and "thought collective". The "thought style" is the collection of concepts, ideas, observations, hypotheses and theories which are prevalent at a certain point in time. The community, or the "thought collective" as Fleck calls it, sees reality through this network of information. The people using the thought style are active in their interpretation of observations and results, that is, during the observation process, they use the lens of the thought style and thought collective and influence the way they "see" the results.

For Fleck, the distinction between different thought styles and thought collectives isn't as clear-cut as it is with Kuhn's distinction between different paradigms. A researcher can use different languages of different paradigms and shift from one collective to another and back again - just as a physicist today can use Newtonian language in one situation and relativistic or quantum language in another.

If so, how does Fleck define the process of scientific discovery?

How does he view the process which Kuhn described as a revolution, in which a transition takes place between one paradigm and another?

Fleck talks about a process which is much more complex, he describes flowing from one "thought style" to another "thought style". This is a fluid and organic process which isn't rigorously defined, and which doesn't consist of discrete, clear-cut stages. From Fleck the process begins with the obtainment of results which don't adhere to the thought style's clear framework. When these results or anomalies begin being consistent, the community begins suspecting that this isn't some arbitrary matter and that an explanation is needed. This unclear feeling, in which the researcher feels like he's heading in a certain direction but can't explain his findings, Fleck calls "resistance". The collective is what creates the resistance as a new phenomenon and defines it as a new concept. When resistance is stable and consistent, the researcher and collective begin creating a new language and new concepts which will explain their feeling, even if they haven't explicitly identified it yet. The better explained and well-defined the resistance is, the more consistent and stable it becomes, until it becomes a scientific idea in its own right, which is part of a new "thought style". So, Kuhn's revolution is the development of a thought style in a process of gradual transition from one view to another, through intermediate stages in which thought styles co-exist and the boundaries between them aren't easily defined.

The views of Kuhn and Fleck, especially the latter, are highly reminiscent of the insights we've gotten to know when discussing network and complexity theory. The scientific community is a network of connections between people, concepts, methodologies, technical equipment, hypotheses, theories, and so on. This network effects the interpretation of findings in a way that's dynamic and active. As part of the network of scientific activity, a variety of results and hypotheses appear. When anomalies and a lack of compatibility appear, they introduce first signs of chaos which can lead to a crisis. From this crisis, and as the anomalies grow stronger, creating a stronger resistance, the community arrives at the edge of chaos, the result of which is the emergent appearance of a new scientific fact, along with a new paradigm or thought style. The network doesn't collapse or change in an instant, in some lightning-fast revolution, but rather it develops into the next stage in its order and internal definition regarding the relations between all of its components.

Will the physics of the concept of time, in the second half of the 21st century, be on the edge of chaos, leading it to the appearance of a new emergent theory?

The revolution's potential - the physics crisis of the 21st century

I am not a fool and so I don't plan on dealing in prophecy. As we've already understood, scientific development and paradigm shifts are complex processes which combines results of experiments, various hypotheses, "good sense", an entire community of vested interests, budgets and a world of concepts, and ultimately, it's people who make us transition from one paradigm to the next. We can't prophesy the development of such a complex system, but we can identify trends, examine the development process and try and identify the various stages the system is in, and the directions in which it can develop.

Is contemporary physics, in the context of the way the scientific community views the concept of time, in a stage which can indicate a coming change? In other words, is the existing paradigm regarding the concept of time about to undergo a revolution? I believe the answer is yes. The timing of this move can be disputed, and it may last more than a century, but I believe that the situation we're in tells us that after 2,500 years, something deeper has got to change.

At the end of this chapter I will try and describe the anomalies and difficulties which the paradigm faces today, the conceptual structures presented in the attempt to deal with these questions and the "ad hoc" solutions raised within the paradigm. Later on, I will outline the foundations of a possible solution which could serve as a replacement for the paradigm.

Even those who feel safe and secure with where the paradigm is at today, and who treat this discussion as a pointless philosophical and metaphysical discussion, convinced that physics has a well-established, well-founded view of the concept of time - have to admit that within the physical community itself, there are still contradictory opinions when it comes to time. In the previous chapter we've seen two esteemed physicists who spent their entire lives studying the question of time, arriving at completely opposite conclusions, which can be ascribed to the first two people having this discussion: Parmenides and Heraclitus. The fact that this discussion is still taking place within the physical community requires some explaining, at the very least. I think that the problem is deeper still: I see a complete lack of agreement on the most fundamental level when it comes to the understanding of one of the most basic concepts of the existing physical theory. While this theory provides us with practical tools for leaning about and understanding the world, we're operating on shaky foundations whose weakness, in my opinion, is responsible for some of the biggest riddles the paradigm is facing today.

The general public perceives science as it is today as having the ability to provide a complete explanation of physical reality, but to be honest - this description is very far removed from reality. There are large gaps between theory and observations. There are impressive theoretical and mathematical structures which describe the structure of the universe and explain the phenomena we find in it, which have no expression in observations of physical reality, and there are observations which raise difficult philosophical questions which remain unsolved. Dark matter and dark energy, for example, are conceptual structures devised in order to fill in the gaps between theory and observations without having any experimental evidence. No one has ever seen dark matter or dark energy or parallel universes or a cat that's both dead and alive at the same time. Physical theory may be highly impressive and helps observe the world with a great deal of success, and it also has practical applications in technology, but this doesn't mean it's complete, limitless and undisputedly founded.

The concept of time isn't understood by us. Is it a physical entity with existence or merely an illusion of our consciousness? Is time an additional fixed dimension, or is it dynamic and evolving from one moment to the next? Does it have one direction or is that just our subjective experience? Does its existence predate the rest of the world's phenomena, or is it emergent, appearing from something even more fundamental?

Does the fact that we have no answers to these questions mean that the paradigm is in a crisis? Some will claim that there is no paradigm when it comes to the concept of time, but that's not exact. Relativity theory defined time in a very specific way 100 years ago. The problem doesn't lie in the practical use of the concept of time, as defined in the framework of special and general relativity, but in the gap between the physical concept and the way it is reflected in our everyday observations of the world. In other words, while relativity theory is good at explaining the relativistic nature of time, and its predictions perfectly match our observations of the universe, the theory doesn't match the uniqueness of the present with its dynamic, developing nature we experience at each moment, and the various degrees of freedom that we have as agents with free choice, operating in the universe within the framework of the time dimension.

Furthermore, the gap between the timeless mathematical structures and formulas, which are the basis of the theory, and the dynamic and changing experience of each and every moment which isn't expressed in these mathematical structures requires an explanation - and so far, no good explanation has been found.

Physics' inability to combine relativity theory and quantum theory into one "theory of everything", is believed by some researchers to be a result of the current prevailing paradigm regarding time, and in order to solve this lack of compatibility, we need a different description of physical time. As I've already mentioned, Karl Kusher's statement regarding the problem of connecting relativity theory and quantum theory and its ascription to the opposite ways in which they view the concept of time was very influential on the community as a whole, and lead Julian Barbour to focus on the problem of time and try and solve it.

The unsolved issues which are raised by the measurement problem in quantum theory and the question of the arrow of time have led physics to develop controversial structures involving multiple universes existing simultaneously, and this is precisely an attempt to deal with the problem of time using "ad hoc" solutions which at present do not meet the scientific principle of falsifiability. The multiverse theory is an excellent example of a paradigm's attempt at continued survival, and I think it counts as a significant anomaly which could indicate the need for a new paradigm.

To summarize, when we examine the lack of compatibility and the discussion on the question of the "block universe" which stems from relativity theory and its implications on free will; quantum theory with its great riddles regarding the measurement problem and the significance of quantum entanglement which remain unsolved; the question of the arrow of time and its physical reversibility; the difficulty in combining relativity theory and quantum theory; and ultimately the large gap between mathematical equations and our experience of time in the universe - we find a situation which definitely sounds like a crisis. Is this crisis solvable or is it a doorway to another revolution? This is a question we can't answer at this stage, but it's not a question that can be ignored either. We need to keep on working on the revolutionary science taking place at the community's margins, which may one day lead to a new and interesting proposal.

Lee Smolin summarizes the issue in his book:

"All of the mysteries physicists and cosmologists face – from the big bang to the future of the universe, from the puzzles of quantum physics to the unification of the forces and particles – come down to the nature of time."

CHAPTER 14:

The Paradigm's Margins - New Insights in the Second Decade of the 21st Century

Preparing for a new paradigm - Moving on to the next step

Now is the time for revolutionary science. It is the time for a new path.

My personal journey in the study of time began in 1999 ("late 20th century" sounds better). Regardless of the closing of the millennium and following my academic studies, I realized to my surprise that what I used to take for granted is now a formidable riddle. As we mentioned above, time is one of the most mysterious of the fundamental concepts. To this day, time still hasn't received an explanation which can reconcile the three worlds which deal with this concept: physics, philosophy, and human intuition. My personal journey in the study of time led me to fascinating and mysterious places, to the big questions mankind has been struggling with for as long as we can remember, and to the mystery that just keeps on increasing as science uncovers more layers of physical reality - a mystery revealing itself both at the widest level of the entire cosmos and at the level of the fundamental building blocks which make up everything. The fact that we've gone through so many different and surprising stops during this journey teaches us that ultimately, in a way, everything's connected: space and time, the primary components of matter, the origin and development of the universe, the mind, consciousness and life itself. Along the way, we began identifying interesting hints which could testify to this connection. I believe that our intuitive perception of time and the way in which it flows and creates a uniform, developing story is the central testimony to the fact that there's a mysterious link between all of the universe's components. Now my journey to understand time is coming together with the journey we've gone through in this book, pursuing this mystery. We passed by some of the stops in this journey with ease, and some of them required us to dedicate time and effort in order to understand where we are, and where we go from there. It may be that some of you feel that despite the great distance we've covered, the view from the window at the last stop is pretty similar to the one we were looking at when we set out. However, I believe that this feeling is far from the truth. The truth is that this collection of information before us is of immense value - not in terms of each separate field of knowledge, but in the connection of all these separate worlds into one, complex system, from which we will receive an emergent answer which will lead us to the development of the next order in the human understanding of the concept of time.

In order to pave a new path of thought, we must break patterns, undermine foundations and combine entirely new ingredients into the stew that is the concept of time. I believe that as long as we keep thinking and talking within the conceptual framework that as accompanied us, being copied from one paradigm to another for 350 years, we will not be able to solve some of the big questions, and we'll keep treading the same paths as we try and form a new figure which will embody both Parmenides and Heraclitus. The time has come for a new concept of time, which will match the predictions of physics on one hand and will be compatible with our strong intuition regarding the dynamic nature of time and the universe, on the other.

For this purpose, I will attempt to link between all the components of knowledge we've encountered so far and create a network of nodes from which a new emergent insight will arise, which will lead to the much-desired phase transition. In this chapter I will leave the comfort zone that is the physical and philosophical description of the past 2500 years and lead you towards unknown lands. Some of the theories and arguments I will provide here have been published by highly esteemed physicists during the past few years, some of them are based on a combination of existing theories from different fields which haven't been unified into one description yet, but which I believe should be so combined, and other arguments are far-reaching hypotheses based on new directions of thought I'm proposing. As I describe my ideas, I will try and provide as much foundations for these arguments as possible, and distinguish between what is considered normal science, according to the majority opinion, and what's considered revolutionary science, the result of physicists of high standing in the scientific community, and what constitutes entirely new directions of thinking.

The universe as a complex system of "nows"

At the start of the 21st century, Julian Barbour introduced us to the view that claims there's an infinite set of possible configurations/"nows". Could time be a product of this set of configurations?

A configuration is a frozen image of all the matter in the universe at a certain moment, in the particular manner in which it's arranged. Each moment in time is a certain configuration of the universe which has been realized, among a set of possible configurations which haven't all been realized in reality. There may be configurations in which you are presently reading this book, configurations in which there is no life in the universe and others where there are no stars or galaxies. In some configurations I finish writing the final chapter and in others I give up on the whole thing at the last moment, found a start-up and retire at 50 (or in other words - "delusional"). Everything is possible. Any possible arrangement of the matter in the universe can appear in some configuration.

Each configuration, each such momentary image, does not exist in reality but has the potential of becoming realized in reality or it might be better to say that the configurations have a different level of existence. The set of configurations is in a superposition: they all exist and don't exist simultaneously. Just as a particle doesn't have certain properties at a certain moment, but rather its wave equation describes the probabilities of it receiving certain properties once a measurement is performed on it - in that same way, each configuration has no full existence in reality, and the probability for it to be realized in reality is defined by the wave equation of the entire universe. A configuration becoming "realized in reality" is our experience of what happened in practice, in contrast to what hasn't. If, for example, there were only two possible configurations, one in which you're reading this book at this moment and one where you're on vacation in the Caribbean, then the configuration which was realized in reality at this moment is the one in which you're reading this book, and the second configuration is now guaranteed to never become realized. The reason for why it was this configuration which became realized in reality rather than another one is that its probability of becoming realized in reality was most likely higher than that of other configurations.

It's possible that this description is aligned with the presentist view, which grants a special status to the present moment, in terms of existence, compared to moments in the past and future. That is, the property of existence can be ascribed to moments in the past that have already become realized in reality, and the present is the transition in which an additional moment receives the property of existence. The future has no complete property of existence yet, it consists solely of potential frozen moments. In other words, it is a combination of the "block universe" view and the view which accords a special status to the present: moments in the future exist to a certain extent, just as particles "exist" before we observe them, but this existence is incomplete and is merely existence in superposition. Realization in reality will grant this moment a more complete status of existence, compared to future moments. The moments of the past already have complete existence, while the present has an additional property which is the act of realization. This is the "collapse" of the wave equation, which grants existence to a certain configuration in a certain moment, rather than some other configuration. The act of realization is unique to the present moment, and once it happens that moment receives an existence which is identical to that of all past moments. For this reason I see this view as a kind of intermediate view between absolute eternalism with its collection of frozen moments with identical existence regardless of their status as past, present or future moments, as the "block universe" theory describes, and the view of presentism which ascribes existence solely to the present moment, in contradiction to relativity theory. In my humble opinion, such an extension of Barbour's approach allows us to describe the universe in accordance with relativity theory and the "block universe" implied by it, since ultimately, all moments exist in some way or other, while enabling us to accord a certain uniqueness to the present moment, and to reconcile our intuitive experience of the uniqueness of the present moment, and the sense that the future is "open", rather than pre-determined.

ER=EPR - Entangled worm holes

The second part of our puzzle is a hypothesis which can be placed on the boundary of revolutionary science. It is the product of joint work by two of physics' most interesting figures in recent years, whose joint research regarding black holes has produced a proposal which constitutes a conceptual breakthrough which could serve as the foundation for the combination of the three great theories (relativity, quantum, and complexity).

Leonard Susskind, a veteran physicist from Stanford who is considered to be one of the fathers of string theory (a physical theory which developed as one of the attempts to build the "the theory of everything", which tries to explain reality using a fundamental unit which is the origin of everything - the "string"), received a strange message from Juan Maldacena one day, a young physicist from Princeton. Maldacena is considered to be a rising star in the world of theoretical physics, and in recent years has been Susskind's partner in promoting the idea he summarized in that message using a single equation: ER = EPR.

I've already described the EPR Paradox at length in the chapter on quantum entanglement. As you recall, the people who came up with the paradox showed that in the framework of quantum theory, several particles (or quantum systems) can have a certain connection between them and affect each other directly and immediately without any exchange of information between them. Meaning that the influence takes place at a speed faster than the speed of light, which is the maximum speed in nature, and so this phenomenon contradicts the property of locality in the world. In other words, this paradox contradicts our knowledge that no body can affect another body without any physical interaction or transmission of information between the two. The EPR Paradox is essentially the pioneer in the discussion on this "strange" connection we find in quantum systems, which led to one of the most important and researched concepts in recent years: quantum entanglement. I've also mentioned the initialism "ER" in the chapter on "worm holes", when I presented "Einstein-Rosen bridges", also known as "ER bridges". Of course, these are the same Albert Einstein and Nathan Rosen of the EPR Paradox, but they developed the two concepts separately, without linking between the two. The "Einstein-Rosen bridge" is a hypothesis of theirs which is based on the equations of relativity theory, according to which, when space and time become significantly curved, a bridge is created between two points which are distant in space-time, theoretically allowing for immediate travel from one point to another despite the great distance between them (chapter 6 includes a more detailed description of this hypothesis). Space-time can become significantly curved due to a large mass or energy (such as in the case of a black hole), and a situation is created in which two points which seem very distant to the observer are actually connected.

Einstein and Rosen published their articles on these ideas in the same year. Despite the proximity of these publications, we don't know that they made any links between these two hypotheses. This link was only made in the past 2 years by Maldacena and Susskind. In the context of black holes, the two claim that there's a connection between the entanglement of two black holes and the creation of "worm holes" - that is, the creation of this type of quantum connection between two black holes creates "Einstein-Rosen bridges" between them. If we take two entangled black holes and convince Bob to jump into one black hole at a certain point in space-time, and Alice to jump into the other black hole which is at a faraway point in space-time, theoretically they would meet each other instantly (even though in practice they would not survive the jump), since black holes are connected to one another through a "worm hole" which constitutes a kind of tunnel or bridge between these two points. According to Susskind and Maldacena, the quantum entanglement between the two black holes is the cause of the creation of worm holes between them. In a way, which we don't understand at present, the entanglement between the two black holes, this "strange" quantum connection, creates a physical object in the universe called a "worm hole".

This idea was part of a wider move aimed at understanding what space-time is made of, and what goes on inside a black hole. The studies and discussions on these questions have produced various hypotheses and interesting directions of thought, such as research branches in string theory and the holographic principle - the idea that the reality we see is actually a holographic illusion stemming from information kept in a world with one dimension less than ours, the way a hologram is an illusion of a three-dimensional image which results from information stored in a two-dimensional image.

These directions of thought and research have led Mark Van Raamsdonk from the University of British Columbia to publish an article in 2010, discussing the relation between the structure of space-time and quantum entanglement. In this article, Van Raamsdonk raises the idea that quantum entanglement is the "glue" which connects the parts of space-time, that is, in some way, quantum entanglement can explain certain properties of space-time and its uniform fabric.

Based on these ideas, and following his joint work with Susskind, in 2013 Maldacena published an article called "Entanglement and the Geometry of Spacetime". In this article Maldacena gives an affirmative answer to the question he presents in the article's introduction: does the strange phenomenon of entanglement in quantum mechanics create worm holes connecting between distant points in space?

Subsequently, today there is serious discussion and research dedicated to the claim that what makes space-time a smooth fabric is the quantum entanglement between particles in different areas of space-time. This link between quantum entanglement and the geometry of space-time stirs a good deal of excitement and interest among the physical community, but it's still in very early stages and must ripen significantly in order to receive a more well-founded status.

In order to understand this idea, we will use an analogy to a sheet of cloth. As we know, cloth is made from threads that are interwoven, creating one sheet with a uniform fabric. The sheet can bend and fold but its basic structure remains fixed and uniform. If we separate the stitches or unravel the threads, at first, we will get separate sheets of the same cloth, and later on we will be left with nothing but threads. According to Maldacena and Susskind's approach, space-time is a kind of "tapestry" created by interweaving "threads" in so that they create one, smooth sheet. These threads are quantum entanglement. They're the ones keeping everything together and making the properties of the overall arrangement different to those of the individual components.

I glean two important insights from this discussion. The first insight is that nowadays there's significant evidence and well-founded arguments which lead us to the understanding that quantum entanglement creates a link between individual, separated components in the physical universe. While quantum entanglement does not transmit information immediately in a way that contradicts relativity theory and the consequent principle regarding the highest possible speed of light in nature, the entanglement creates a certain connection between the entangled components which allows us to examine them as though they comprised a single system with new properties. The second insight is that there is probably a deep and basic link between quantum entanglement and the properties of space and time in the universe. That is, quantum theory expresses something which is more basic, essential and significant than a description of the mechanics between particles in the universe, and so the principle of entanglement and quantum theory as a whole must be a basic and central component in the description of reality, including the description of the concept of time.

So, in the discussion on the fabric of space-time, we see the appearance of concepts and terms which are reminiscent of the world view we find in network and complexity theory. While we still haven't seen a clear and direct link between these two worlds, when we talk about an entity in which connections between a collection of individual components create new properties which did not exist in the components themselves, we're using the language of networks and complexity. We will soon see that the transition to a richer language and a clearer connection between these two worlds is just a matter of time.

And indeed, in the margins of the physics world, general statements can be heard, which combine the concepts and language of networks and complexity, and they hint at a possible developing connection between complexity and cosmology.

In one of his lectures on the entanglement of black holes, Susskind himself claims that even though there is no worm hole between two particles the way we'd expect to see in the case of two black holes, he believes that the model he presents in the context of black holes also describes particles in a state of quantum entanglement: "It seems as though there are microscopic super-planckian worm holes between entangled particles".

He doesn't think this fact has any special significance, and on this point I disagree. One of the goals of this chapter is to try and explain how this view could be expanded into a wider and more comprehensive view of reality.

Furthermore, in another lecture he claims that it's time for complexity theory, which has mostly been expressed in computer science and biology, to find its way into cosmological physics: "complexity theory has not fulfilled a significant role in physics. I believe it's going to be lot more significant in the future."

Network infrastructure - Space-time as a network

When we got to know the world of complexity, I told you about systems comprised of a large collection of components whose interaction cannot be accurately predicted. Now imagine the complexity of an object comprised of interactions between billions of particles. A very small object can consist of several billion atoms whose joint activity results in the emergence of new phenomena which aren't found in the atoms themselves, such as temperature, color, conductivity, and so on. The number of interactions of a small collection of atoms can be astronomical, and when we're talking about the entire universe, the number of components and interactions are practically infinite. No system could be more complex than this. Anyone who still thinks that the universe is a machine and that explaining it requires merely arriving at some Newtonian, mechanistic formulas, will always find himself reaching a dead end.

The only way to handle such a system is using insights from complexity theory, and indeed, Brian Swingle from Stanford University uses network language to describe the geometry of space-time. Swingle proposes making use of network models in order to understand space-time, and examining this complex system not through its components, but rather through a comprehensive view, monitoring the system's development from some starting point. The mathematical aspect of this proposal isn't important for our purposes at the moment; what's important is that Swingle is proposing that space-time consists of a sequence of nodes connected in a complex network, and that each node consists of entangled particles. In other words, following the discussion of Van Raamsdonk, Susskind, and Maldacena, space-time should be viewed as a network in which each node can be a particle or a pair of entangled particles, and these nodes are connected via quantum entanglement. The structure of this network is the tapestry called "space-time". From this network of entangled particles, we get the properties of space-time as we know it - a smooth, uniform sheet. So space-time is emergent from the network interaction of entangled particles.

At this point, we should ask a more fundamental question: if the network geometry of the universe is the infrastructure of space-time, which is the universe's most fundamental fabric, what does this mean regarding the network model? Should we treat the network model as having a basic significance in our universe, and look for it in other fundamental components as well? It's possible that we need to look at the world through the lens of network theory, examining physical reality as one based on a network pattern, and which operates according to the conceptual and procedural view of the world of networks.

I believe that today, more than ever, this point of view could help us solve some of the greatest riddles which have accompanied mankind, and which I have described along our journey in this book. Could time be an emergent phenomenon of complex system, built as a network in reality? Could time be a product of this network of "moments"?

The network and connectivity - quantum entanglement

I will now present a possible expansion of the ideas described so far in this chapter. It is a new direction of thought which to date, hasn't been proposed in the physical or philosophical community.

In order for a network to be created, we need connections between the various nodes. These connections could be power lines in a power grid, links between websites or synapses created between neurons in the brain. Following the ER=EPR idea, and the network view of space-time, Brian Swingle proposed the possibility that the links in the space-time network are what we call quantum entanglement. I will now present the possibility that the nodes in the universe-network, connected via quantum entanglement, are the configurations (the "nows") we encountered in previous chapters. Quantum theory teaches us that two particles or systems can be connected to one another in a special way, which isn't clearly understood, regardless of the physical distance between them. Accordingly, we can claim that the different configurations in the space-time network are connected to one another via the quantum entanglement of the particles which make up these configurations. But not every particle in any configuration is entangled with any other particle in any other configuration - there is a certain regularity which determines which particles get entangled. This regularity has to do with the similarity in the particle states in different configurations. Each configuration includes within it different particles which are entangled with one another. If there is quantum entanglement between two particles in a particular configuration, and those particles are also entangled in another configuration, a connection between these configurations is created. So, the bigger the similarity in terms of quantum entanglement between the particles which makes up different configurations, the more connections are formed between these configurations.

In order to understand this, we will examine the case of a table placed before us. This table, much like any other material object in the universe, consists of a collection of particles with certain properties, arranged in a certain way, and these particles are entangled with one another in a certain arrangement. In a particular configuration, we get the table as it appears to us now, in other configurations this table could be identical and in other configurations these particles which make up the table could be arranged in a way that yields different properties, and so in these configurations we would not see this particular table. According to the hypothesis I'm proposing here, when particles are in a particular state in different configurations, entangled with one another in each configuration in an identical way - such as when the table comprised by these particles is found in different configurations - then the correspondence between the entangled particles in each configuration creates a connection between the configurations themselves. This means that the number of connections between every two configurations in which the arrangement of matter is similar, is larger compared to when the number of connections between every two configurations in which the arrangement of matter is different; when the arrangement is different, we will find less cases of identical entanglement and therefore less connections between the configurations.

So, in this world view, we get a network of configurations connected among themselves via quantum entanglement, in accordance with the similarity between the entanglements existing between the particles which make up each configuration. From this, we can conclude that the bigger the number of connections between two configurations, the greater the similarity between the arrangement of matter in each configuration.

In the reality we experience, the transition from one moment to another isn't felt as a "leap" between states, which means that the arrangements of matter and quantum entanglements in these two moments are very similar. The particles entangled with one another at the present moment will probably find themselves in the same quantum entanglements in the next moment, except for a minor change; the similarity between the present moment and the next is the highest possible similarity (same goes for the present moment and the previous one). This means that the present moment which becomes realized is maximally connected to the moment before it and the moment after it. There can be no configuration whose similarity to the present one - and therefore the number of connections to the present one - is bigger. It's possible that at this point the direction I'm getting at is becoming clear.

The arrangement of matter in a particular moment also includes the quantum entanglement existing in this moment between the configuration's components. So each configuration is itself a network of connections between the particles which comprise it. The network is comprised of nodes, which are the sub-atomic particles arranged in it, and of the connections between them, which are the internal quantum interaction created by quantum entanglement. This means that each configuration is a network in its own right, which includes many connections between the particles entangled within it. So each node in the universe-network is a possible configuration, a network in its own right with a great deal of complex connections - both internal connections between the particles which comprise it and external connections to the rest of the configurations in the universe. Ultimately, all of the components in the universe are connected in a network of quantum entanglement: a network which connects all the possible moments in time (configurations), in which each such moment is a "ball" of connections between all of the components which make it up.

This network of configurations is visible to an observer who is outside of the universe, as described by the "block universe" and the Wheeler-DeWitt formula: a timeless world of configurations ("moments" in space-time).

This contradiction is precisely the "dead end" we keep arriving at, and we must solve it once and for all, so we don't find ourselves going back to a familiar and frustrating starting-point, alongside Sisyphus, time and time again. We must find a world view which allows for a network of configurations which isn't dynamic in and of itself from the standpoint of an external observer, as is implied by relativity theory and the "block universe" on one hand, but which has an internal dynamic nature which makes the development of each configuration and the network as a whole possible, on the other.

As I've described previously, the fundamental essence of the network as a whole is summed up by the quantum entanglement between its components. This entanglement is what constructs the internal connections inside each configuration and the connections between the various configurations. In order to explain the network's dynamic nature and development we must explain the change and development of the network's connections - that is, explain the dynamic nature of quantum entanglement. The dynamic nature of quantum entanglement affects the network's development and it is the missing, essential dynamic component in the description of the universe and reality.

The "collapse" of the configuration - The realization of the "now"

As we've seen in the chapter on network systems, networks which have the capability of creating new structures and emergent phenomena have an additional important property, which is "Preferential Attachment" or, "winner takes all", which creates the structure of a scale-free network. The network of configurations is a scale-free network since similar configurations will probably be more highly connected. Why? As we mentioned above, the configuration's arrangement of matter is supposed to have a certain influence, if not a primary one, on the entanglements which will eventually develop in that configuration, and so if matter is arranged in a similar way in two configurations, similar entanglements will develop in both of them. If the connectivity between configurations is based on the level of correspondence in their internal entanglements, then two similar configurations will be more connected to one another.

When a particular configuration is very highly similar to other configurations in terms of their entangled particles, it has many connections with other configurations and it can become a hub of connectivity in the network. That is, it can become a hub with a higher connectivity than other configurations, in accordance with the "winner takes all" rule. As the network develops, additional configurations which are similar to configurations connected to these hubs will join the network and receive a large amount of connections to other configurations, since any connection connecting between two nodes also connects each node's set of connections. For this reason, any additional connection exponentially increases the connectivity level of each such node in the entire network.

Now that we've understood how the network grows and develops, we can begin proposing possible answers to the question which is the reason we've all gathered here: how is the thing we call "the progression of time", or the experience of shifting from one moment to another, created from the network structure of these configurations?

What we've described so far is a network of configurations, that is, a network of static arrangements, which all have the potential of being a "moment" in time. Each such configuration could be a single image in the "film strip" of the universe. However, in this network, configurations exist in a superposition, which means they only exist potentially; not every such configuration gets realized in reality, and the continuity of time we experience includes just a small part of these potential configurations. What then, is the process which causes certain configurations to become realized in reality, and why do only some configurations become realized in reality, with others remaining merely potential configurations?

In order to answer this question, we will need to return to one of the interpretations for the measurement problem I presented in the chapter on quantum theory. As you recall, following the measurement problem and the question of why the wave equation collapses into one result out of a set of probabilities, various interpretations of quantum theory were proposed, whose goal was to add additional elements or explanations in order to complete the picture and extract the mysterious, "magical" element from it. I haven't described all of the different interpretations of quantum theory which appeared over the years, but rather I focused on those which could have significance in our journey to understand the concept of time.

One of the interpretations I presented is the Ghirardi–Rimini–Weber interpretation (GRW theory), which aims to solve the measurement problem and explain why we **don't** see the phenomenon called superposition in the macroscopic world. According to the GRW interpretation, every particle spontaneously collapses at some point from a state of superposition to a specific, defined state. The de-coherence phenomenon explains why the collapse of one particle leads to the collapse of the particles around it, and they leave their state of superposition. When we look at a macroscopic object (such as a table), we don't see it in a state of superposition since it consists of a huge collection of particles, and so the probability of one of them spontaneously collapsing and causing the other particles to collapse into a defined object which isn't in a state of superposition is very high, virtually 100%. Since all objects in the macroscopic world consist of a large number of particles, the probability of us seeing a macroscopic object in a state of superposition is so low, it's practically impossible.

Now we will try and apply the GRW hypothesis' principles to the way a particular configuration in the universe-network becomes realized in reality. In order to do so, we will combine this view with the "phase transition" process we encountered in the context of network theory: a network's development to the point in which it reaches a certain threshold and emergent properties and structures are created within it. If we return to our world of configurations, and in accordance with the GRW theory, the more quantum entanglements take place within a configuration - as consequently between it and other configurations - more connections are created in a network; the configuration eventually arrives at a level of connectivity which crosses the system's collapse-threshold, becomes a hub and collapses into reality. This is how we obtain a collapse of a particular configuration from the superposition of configurations without needing an observer-measurer.

Unlike the GRW interpretation, the collapse of a configuration isn't random and spontaneous, but rather it is a result of the amount of connections a particular configuration has, and the fact that it has become a hub in the network. The more instances of entanglement a configuration has, the more space-time gets created, changing the configuration from a potential configuration in a state of superposition with the other configurations, to a realistic one. In each and every moment, as a result of the synchronization process created between the configurations in the universe, a particular configuration crosses a certain connectivity threshold and the network goes through a phase transition to the next stage. In the phase transition, the configuration collapses and becomes realized in reality in accordance with the amount of entanglement within it.

So, just like a particle described by Schrödinger's wave equation is in a state of superposition until it is measured, at which point it "collapses" into a singular state, becoming realized in reality, so is a "moment" in time in a state of superposition until at some point it crosses a connectivity threshold, "collapses" into reality, causing a phase transition in the entire network. Thus, in each and every moment, one after another, certain configurations become realized in reality. The sequence of configurations becoming realized in reality is like a sequence of images in the film strip of the universe, and it creates an emergent phenomenon called "time".

In principle, each moment could be very different from the one that came before it, since each configuration which gets realized in reality can be very different from the one that came before it. Such a difference between the moments we experience would negate the coherent continuity of time: each moment in reality would "leap" to a different arrangement of matter in space and a very different experience compared to what was experienced just a moment ago, and all the memories and history of previously realized configurations (what we refer to as the "past") would disappear. Such a world would be meaningless, we wouldn't even have the ability of being aware of the fact that each moment is highly different than the last, since the information about the past would not have been identical between one moment and another.

But this isn't the world we know and experience, or the one we'd want to live in, and so in order to rule out such a world we must explain why such leaps do not occur.

So how does the process in which configurations become realized maintain the coherence of the continuity of moments, creating a reality in which each moment is slightly different than the one that came before, including information on all previous moments, leading to the creation of a coherent, continuous and logical story?

First I will touch on the question of coherence - why is each realized configuration only slightly different than the one that came before it? And in the analogy we've been using: why is the sequence of images on a film strip a sequence of highly similar images including very minute changes, creating an illusion of a narrative continuity? The answer to this question lies in the correspondence between the entanglements of the configurations which become realized. As I described earlier, the configuration becoming realized in each moment is the one with the highest connectivity level, a connectivity level which causes the configuration to cross a certain threshold whose result is a "collapse" into reality. The configuration's connectivity level has to do with the correspondence between the entanglements between the particles which comprise it and the entanglements in the configurations connected to it. The correspondence of quantum entanglements between configurations indicates a correspondence in the arrangement of matter between configurations. In other words, according to the "winner takes all" approach in network theory, if there is a correspondence between the entanglement existing in a particular configuration and the entanglement found in the last configuration to collapse, this means that the synchronization between those two is the highest possible one, and so the probability of it becoming realized in reality right after it becomes almost certain. So, if each and every moment involves a configuration becoming realized in reality as a result of its high connectivity with other configurations, including the previous configuration, there's a high probability of the arrangement of matter between one moment and the next being highly similar, including relatively minute changes. Just as in the continuity of images in an animated movie - each image is almost identical to the one that came before, including just a small change which is only noticeable when we're experiencing the entire sequence of images. Similarly, configurations becoming realized in reality are a sequence of near-identical images with slight changes between them, and experiencing them in sequence creates a continuous, consistent and coherent story.

The second question I raised has to do with our information about the past which goes on accumulating from one moment to another. How do we keep the information about all of history from the dawn of time until today in a continuous manner, and how does this information keep on accumulating and developing following every moment that gets realized? As you recall, in order to answer this question, Julian Barbour created a theoretical structure called a "time capsule". According to Barbour there is no need to accumulate information, or have it pass from one moment to another, it simply exists in each and every configuration. Any image includes information about its past, the change that came before it and of which it is a part. Imagine yourselves standing in the street and all of a sudden, you're given the supernatural ability to stop time (I'm convinced that every one of you has wished for this ability at least once in your life). Now stop time. Look around you: traffic has stopped, but this image has enough information for you to know which car is going in which direction, which pedestrians are standing and which are walking, whether it's raining or the sun is shining, and which direction the wind is blowing. You might find a diary including information about the past of the person who write it, and maybe even find a history book which will recount all of human history. So a static image can include an infinite amount of information about its "past". Does the presence of such information mean that this image has a "past"? Certainly not. It's possible that the person who painted this image included all of this information in it and stamped this fictitious "past" into the image. A history book with all its pages and letters, same as brain cells including memories of the personal past of each and every one of us, or footprints in the sand behind us, are all ultimately a certain arrangement or configuration of matter in the universe. Matter arranged in a way that allows us to draw conclusions, but this doesn't mean that this image has a past. So, the fact that each moment has information about the "past" and the fact that each moment is similar to the one that came before, can explain why there is a historical "continuity" between moments and an "accumulation" of information about the past.

Now the final and hardest question of them all arises: how do configurations and the network as a whole develop in a timeless world? If the universe's fundamental components are timeless, and so is the network comprised by them, where does our experience of the duration of time come from?

The internal clock - Emergent time in entanglement

Heat, the ability to plan and manage in a swarm of ants, life and consciousness - the network and complexity theory of the 21st century has taught us that all of these are emergent phenomena which only appear from a combination of a high number of components interacting with one another. What's interesting is that these emergent phenomena include a real experience of existence for us, but they do not exist in the components which comprise the system, but rather only in their overall interaction. This definition essentially means that these phenomena are an illusion created from the overall system, but each of these phenomena and many other similar ones are very real for us. Is time a similar phenomenon? Is time a real phenomenon while ultimately being merely an emergent one?

As was made clear in previous chapters, the physics of today - both relativity theory and quantum mechanics - hints at the fact that what we experience as the duration of time could very well be an illusion. The attempt to combine relativity theory and quantum mechanics has led to the Wheeler-DeWitt equation, which also describes a timeless world.

Of course, this conclusion is the subject of much criticism, both physical and philosophical, and many attempts have been made to include the component of time in the overall description of the universe, with no success so far. One of the biggest supporters of the existence of time and its necessity in the physical description is Lee Smolin. Simply put, he claims that everything we've been doing so far stems from mistakes in our understanding of reality and in the fault he calls "physics in a box"; we try and draw conclusions about the universe as a whole from experiments involving small, closed systems. According to him, applying a conclusion made in a box to the universe as a whole isn't a valid move, neither scientifically nor logically, and so we encounter paradoxes and a lack of consistency when it comes to the concept of time.

However, so far everything points to the fact that we have a problem with the concept of time.

The attempt to understand the origin of our sense of time and the gap between it and scientific theory isn't new. As early as 1983, William Waters and Don Page, two theoretical physicists from Canada and the US, suggested that quantum entanglement could be the source of the emergence of the time we experience in the world and the solution to the problem of time which results from the Wheeler-DeWitt equation. They provided a mathematical proof for the fact that if a "clock" is quantumly entangled with the entire universe, an observer who is part of the universe itself (us, for example) would see the clock ticking and progressing in time, but for a viewer located outside of the universe, the clock would seem to be standing still. In other words, for us observers, placed inside the entangled universe, the change in the world seems like the ticking of a clock which indicates the progression of time, but to an external observer, the world would seem like a collection of moments which are frozen in time.

This point of view is reminiscent of theological and mystical world views we've already mentioned, which place God in the role of the external observer who sees the world as an infinite, static collection of moments, and who doesn't experience the world in the way we do, as progressing from one moment to another. But from the standpoint of Waters and Page, a theoretical observer or the role of the "observer outside of the universe" doesn't have to be filled by some entity; the discussion is about understanding the universe itself, not proving the existence of entities which are external to it. Manning the position of external observer is a legitimate move, but it is based on faith and has nothing to do with the physical and philosophical argument.

However, Waters and Page's argument was a mathematical

thought experiment, based on the Wheeler-DeWitt equation which was known at the time, and the two didn't offer a way of empirically examining the argument and proving it using a system in the universe itself.

In 2013, a team of physicists in Torino, Italy, finally succeeded in carrying out an experiment which could prove Waters and Page's claims. They used a "dummy universe", consisting of just two photons, quantumly entangled with one another. In one version of the experiment, one photon served as a ticking internal "clock" inside the universe, which allows us to examine the progression of time in the system, and the development of the system as a whole on this timeline. That is, the "universe" in this version was entangled with the observer, who's observing an internal clock in the "universe" and essentially performs a measurement inside the "universe". In another version of the experiment, the researchers made use of a clock that was outside the system, and examined the development of this "universe" (which as we mentioned above consists of just two photons) in the timeline of a clock that isn't entangled with the system. So, the observer in this version performs the measurement on the clock which is outside the "universe" and so isn't a part of the universe. The experiment's purpose was to examine how the universe develops from the point of view of an internal observer watching it from "the inside", and how the universe develops from the point of view of an observer watching from "the outside". This is essentially an attempt to empirically prove the gap between our experience of time in the universe and the physical description of a static universe found in the Wheeler-DeWitt equation. The results didn't surprise the physical community which supports a timeless paradigm but would sound like science fiction to anyone else. In the version where the clock was inside the system, photons were observed developing and changing their properties along the axis of time - that is, a universe with time, changing along this axis of time. In the version where the observer examined the system using an external clock, the system was observed as being static and unchanging, i.e. a universe which doesn't include the axis of time. A frozen, unchanging image.

The experiment by the scientists from Torino was the first attempt at applying Waters and Page's hypothesis regarding a simple system as opposed to the entire universe, which resulted in many objections. Some claim this experiment proved nothing new but just re-confirmed the predictions of quantum theory. There is some truth to this claim, but I think that the experiment involves an additional, clearer statement regarding the hypothesis that time is emergent from something more fundamental ("quantum entanglement"). Another, stronger criticism, following Smolin's approach, is that this was "physics in a box": you can't take a "universe" consisting of just two photons and apply the results to the real universe - it is a logical fallacy, and such an experiment doesn't meet the philosophical and methodological standards according to which we're supposed to operate. This is a strong argument and there's no easy way to face it. But this criticism is relevant to any experiment and accepting it could paralyze the ability to conduct empirical experiments in the realm of cosmological research. I believe that this argument isn't enough of a reason to give up on the implications of the Torino experiment, but we must remember it when we attempt to understand the experiment's significance and practical application in a more comprehensive theory.

To summarize, this experiment reinforces the hypothesis that time is merely emergent. It somehow emerges from the complex system known as "the universe" and it is part of the experience of any observer located inside the universe due to the phenomenon of quantum entanglement, but it does not exist for an observer outside the universe.

Is it possible that if we were to look at the universe from the outside, we'd see a collection of configurations, frozen in time? A "pile" of moments without any movement or change?

The next stop - The Network of Time

Are the Torino experiment and the ideas raised by Waters and Page the missing piece of the puzzle? This idea combines the "block universe" which follows from relativity theory and the intuitive experience of progression along a tangible axis of time. We can look at such a universe from two points of view: when looking from outside the universe we see a frozen, static universe which doesn't include an entity called "time"; when looking from inside the universe, an experience of moving through time from one moment to another due to the quantum entanglement between the universe's components and between those components and the internal observer, who experiences time inside the universe. As I claimed at the start of this chapter, the implication is that quantum entanglement has ontological primacy over the phenomenon of time, and functions as its origin.

So we accept a network of configurations, where the connectivity between those configurations keeps on growing due to quantum entanglements within them and between them, and nodes in the network which, when crossing certain thresholds, become hubs and thus cause a phase transition of the entire network and the "collapse" of particular configuration into reality with every moment that passes. The network as a whole appears as a static network described by the Wheeler-DeWitt wave equation, but for the network's internal components, an emergent, tangible time is created from the quantum entanglement between them. Just as "heat" and "consciousness" are very tangible phenomena resulting from the interaction between many components, but which do not exist in any individual component, so is time a very tangible phenomenon which stems from the interaction between many components ("moments" in the network of configurations), but does not exist in each individual moment.

***

The last question we face when describing a static physical world, which has risen in every philosophical discussion on the question of time as long as anyone can remember, is the question of free choice. How can free choice exist in a world with static, pre-determined configurations which become realized in reality one after another as part of a physical process, free choice which is based on the ability to influence the universe's development in some directions rather than others?

In the final chapter we will attempt to provide an answer to this ancient, complicated question as well.

CHAPTER 15:

The Science of Choice - Free Choice in the Network of Time

"All is foreseen, and freedom of choice is granted" - The different approaches

A servant comes to the king, running. He's completely pale. "What happened?" asked the king. "I urgently need one of the palace horses so I can run as far from the city as possible", the servant replied. The king insisted that the servant explain why, and the servant recounted the events of that morning: "I was walking around in the city market and suddenly I saw the angel of death giving me an odd look". The king was astonished and ordered his servant: "In that case, take my fastest horse and run away to the farthest city in the kingdom - Samarra".

The king was angry at the angel of death who had given his servant such a fright and went to the market to talk to him. When he found him among the market stalls, he asked the angel of death to explain his behavior towards the servant. The angel of death was apologetic: "I didn't mean to scare him, I was just surprised to see him as I'm supposed to meet him tomorrow in Samarra...".

This deterministic story, which illustrates that the future is predetermined and cannot be changed, has appeared in many versions since the dawn of time: from Greek tragedies such as "Oedipus Rex" by Sophocles, in which Oedipus' parents tried in vain to prevent the prophecy which states that he will kill his father and marry his mother, to movies in our time which begin with the story's final scene before returning to the start to describe the events leading up to the inevitable ending. The questions of free will and free choice are ancient ones which have been widely discussed in philosophy and science without mankind ever arriving at so much as a single agreed-upon insight, and they are based on similar worlds of knowledge and identical foundations as those of the questions of time and consciousness. Anyone who discusses the question of time eventually arrives at the question of free will, since the answer to the question of whether time exists or is merely an illusion, directly affects the question of free will, and the question of consciousness is of course a fundamental one in regards to free will and could be considered as inseparable from it since will and choice take place in the framework of consciousness. The questions of free will and free choice have deep religious and ethical implications, which is why they've always served as fertile ground for discussion and research in the fields of theology, philosophy and physics, and lately they've begun receiving attention in the field of neuroscience as well. Since these concepts have been discussed within the framework of different paradigms, using different languages, the various arguments created a hodge-podge which doesn't allow us to discuss the essence and existence of free will without running into a dead end which leaves all parties of the discussion entrenched in their views. In this chapter I will attempt to clear things up when it comes to the various arguments and eventually combine them as part of our journey to understand the concept of time.

The various views regarding free will and free choice can be classified into three main groups:

1. **Determinism** - according to this view, the future cannot be changed. All events are predetermined and we, as agents in the world, have no real free will or choice; any desire or choice of ours has been predetermined, or known in advance. Within determinism we can distinguish between several approaches. Causal determinism claims that any event or phenomenon in the world has a cause which precedes it in time, and that when we examine the causal chain we understand that every event, including the event of the formation of free will and the event of deciding to perform a certain action, stems from prior causes which we have no control over. Non-causal determinism, such as assuming the existence of "fate" or the "block universe" view which stems from relativity theory, does not explain the fact that the future is set through prior reasons, but rather claims that all events will happen one way only, due to a fate that guides them, or because they have always existed simultaneously, and it's only us who experience them in "time". This view is somewhat analogous to theological views which view the world through the eyes of some God, for whom all events are simultaneously available, and in which our fate is often predetermined (such as in the concept of predestination in Christian theology). We as humans can be exposed to these future events through the words of prophets who predict the future. Some of the theological views left the door open for free choice for the believer and engaged in heated, contradictory discussions in the vein of "**All is foreseen, and freedom of choice is granted**". "Soft" determinism, in contrast, is less strict when it comes to the limitations which dictate our lives, and emphasizes that within these limitations, we find a certain amount of freedom which allows us to make decisions. While the world may be deterministic to an extent, we can operate freely within the framework of these limitations, even if we can't break through them. And

yet, this approach doesn't define where the line between what is predetermined and what is open to choice passes, and therein lies its weakness.

2. **Libertarianism** - this approach rejects the deterministic component of the world and claims that freedom of will and freedom of choice, whose existence we intuitively accept, must be real. The explanation for our absolute freedom to desire and to choose has to do with various arguments regarding the separation between body and mind, or the attempt to find spaces where free will can be found within the physical realm. So, a libertarian believes that humans have a free will which isn't limited in some way. Its origins could lie in a separate spiritual entity (mind) or physical processes which create consciousness.

3. **Compatibilism** - this view claims that there is no contradiction between determinism and free will. This means that even if the world is deterministic, and the way events unfold is predetermined, this doesn't mean we can't ascribe true freedom to our desires and choices, within the framework of certain rules. This approach attempts to combine the deterministic world as revealed by physics and philosophical views, and the desire to keep the concept of moral responsibility as a factor in the decision-making process behind our choices. Since I believe that this view is a "golden mean" between two contradictory approaches which constitutes a compromise which can't be accepted when building a physical and metaphysical world view, I won't elaborate on this approach in this chapter, and will treat it as a type of "soft" determinism.

Ultimately, the different views can be summed up by a clash between physics and intuition, between body and mind. There is almost no question that our world has limitations which make it deterministic, at least to an extent. Our desire and choices aren't completely free, and mostly you won't find naive libertarians who claim that everything is possible. Physical, philosophical, moral and psychological limitations, among other, impede our ability to choose anything we can think of, and the things we can think of are often determined unconsciously, i.e., in a way that isn't completely free. But our strong intuition of being agents who operate freely in the world, with free will, with a moral obligation to operate in certain ways rather than others, are the strongest arguments against absolute determinism. In order to understand the discussion and the contradictions arising therein, we'll begin with the deterministic description which stems from science, and physics in particular.

Laplace refuses to let go - "Hard" determinism

What does science have to say on the matter? Is the world deterministic and predetermined, or is it our strong intuition regarding free choice and our ability to influence the world that better matches reality?

The scientific revolution which took place in the 17th century has resulted in a completely deterministic scientific world view. Among the thinkers of the time, such as Descartes, Galileo and Newton, an understanding of the world as a machine which operates according to certain rules began to arise. According to the machine's initial conditions and the rules according to which it operates, we can predict with complete certainty what the world will look like in any future moment. The purpose of Newtonian science is a complete understanding of the laws of nature. Once we know all the laws, we can know how the universe works at each and every moment and predict what it will look like at some future point in time. In previous chapters we mentioned Laplace, which provided us with an extreme example of the deterministic implications of Newton's theory, when he described the "omniscient" being which can predict the future with accuracy based on an understanding of the causal relations in nature.

Both the human body and the brain, which is responsible for decision-making and choice of actions, is made from the same particles whose motion, according to Laplace, results from causes that aren't up to them. So if that supreme entity observes all the particles in the universe, including the particles which make up our brain, it'll be able to identify the set of causes starting from the big bang up until today, which have led us to our current state, and explain every choice we made in the past and will make in the future. In a world where we can identify the causes which led us to a certain choice, and in which any future choice of ours can be predicted, there is no room for free will. For this reason, our sense of having free choice is an illusion. The description which follows from Newton's physics is, of course, "hard" and naive determinism. However, from a purely physical point of view, we almost always end up with the same uncompromising description.

Beyond the bothersome contradiction between the physical deterministic description and our strong intuition, there are many complex philosophical arguments which attack the deterministic claim, and one of them is the "vicious circle argument". According to this argument, determinism is sawing off the branch it's sitting on: if the world is deterministic, this means we don't choose to accept the deterministic argument through free choice, but rather, we're being forced to accept it just as we're forced to accept any other claim in a deterministic world. If an argument is forced on us, how can we accept it as real? And if we freely choose to accept the deterministic claim, by doing so we prove that the world isn't deterministic.

Another interesting claim against the deterministic stance comes from a deterministic point of view. If the world is deterministic, and we have no freedom of choice, why did the illusion of free choice develop over the course of evolution? This illusion has no value when it comes to survival, since in a deterministic world, human choices have no value, certainly not when it comes to survival, since humans don't have an actual ability to choose in the first place. So why would the evolutionary process result in an ability that has no survival value? This argument is an example of the paradoxical difficulty that lies in the attempt to logically and scientifically discuss questions of free choice. It's possible that this argument is unfounded since it assumes that evolution is a process which develops freely with time, which can't exist in a deterministic world, meaning that there's no point in discussing the purpose of evolutionary development. The paradox created here is the attempt to discuss processes of development in a deterministic world. The question we should be asking is, why is everything pre-determined in a developing image? Why do we even experience processes of development and learning in a world where everything is pre-determined?

The simple solution is to assume the existence of an additional entity called mind/consciousness/soul, which is entirely separate from the physical world and which can't be explained in scientific terms, which allows us to have free choice. And yet, not only does this solution do nothing to simplify the discussion, it adds components which are far from something we can take for granted, making the discussion much more complex. So, before we discuss the difficult philosophical questions which stem from the existence of a separate entity which doesn't depend on matter, we should try and find a solution within the framework of matter.

The "bleep" leap - "Soft" determinism

Many had hoped that the new sciences which appeared in the 20th century would provide a physical foundation for the solution of the question of free choice, and some still do. But, unfortunately, this hasn't been the result of the previous century's revolutions. Early on, it was clear that Einstein's relativity theory doesn't provide much hope for the matter. In fact, it unequivocally established Einstein's deterministic view of the universe. One of the main implications of relativity theory, which we described at great length in previous chapters, is the "block universe" view, which states that all the events in the universe since the dawn of time exist simultaneously. So, any future event already exists, which means that any choice I make in the future has already been determined and isn't truly free.

Quantum theory had inspired hope among libertarians who thought it would provide the explanation for freedom of will and choice within the framework of physics. This feeling grew stronger among libertarians mostly since quantum theory completely decimated the Newtonian understanding which Laplace had based himself on; namely, our ability to understand all of the universe's components and predict what the universe will be like in the next moment. According to quantum theory, and following the uncertainty principle which follows from it, we can't deterministically predict how the microscopic world, which makes up the universe, will behave like in the next moment. Over the years, many attempts were made to find space within quantum theory in which we could insert the elusive free will, which to date hasn't been incorporated into any scientific world view. A famous supporter of this view is Roger Penrose, who in his book "The Emperor's New Mind", tries to establish consciousness and free will in the gaps left open by quantum theory in the rigid deterministic world view. However, the quantum discussion and its relation to the question of free will are very biased and based on the authors' strong desire to find free components, and even mystical and spiritual ones, in this new science.

Quantum theory has introduced three elements to the deterministic discussion which can be used to find an explanation for the existence of free choice: randomness, probability and the observer's influence on reality.

According to quantum theory, the particles in the universe do not behave deterministically. We can't accurately predict the behavior of each and every particle in any moment in time since the particles' behavior is entirely random in some cases, and probabilistic in others. Reality can be predicted on a statistical level only (for example: there is a 20% chance of a particle behaving in manner X and an 80% chance of it behaving in manner Y). But the fact that the microscopic world isn't deterministic doesn't say anything about the question of free choice, since free choice **cannot** be random or statistical. Free choice should enable the choosing entity to choose whatever it wants without any limitation. We can't choose things randomly since in that case the choice isn't free and intentional, just a random result without any cause which can explain it. In other words, randomness isn't synonymous with choice, and in fact it might even be the opposite of it.

Aside from that, quantum theory describes a probabilistic world in which phenomena cannot be accurately predicted, only probabilities for various events. Does this probabilistic freedom allow us to explain free will? I believe it doesn't. A choice based on statistics is ultimately limited by the statistics themselves, and so is not a free one. If the probability of one possibility taking place is 30% and the probability of another possibility taking place is 70%, this means I'm obligated to choose the first possibility 30% of the time and the other possibility 70% of the time, without being able to decide when I choose each possibility and without being able to change the probabilities. This isn't a definition of free will or free choice. So, the statistical/random opening made by quantum theory doesn't make room for free choice.

The second element of quantum theory - the measurer's influence of the system being measured - is the holy grail of anyone who wants to prove that we have an influence on the universe (a physical, mystical, spiritual, cosmic etc., influence). Many popular views have spread around the world following this approach, and have been expressed in successful movies such as "What the Bleep do we Know?" (2004), and Leap! (2009), which describe mankind's ability to influence reality through the insights of quantum theory. One of the most well-known examples of this phenomenon is the infamous movie "The Secret" from 2006, which drew a connection between quantum theory and western capitalism, according to which if you ask the universe nicely enough you can wake up tomorrow with a fancy jeep in your garage and a million dollars in your bank account (in case you're interested - I tried this and it doesn't really work).

Even if we assume that human consciousness influences the measured result in a quantum system, as Eugene Wigner claims, for example, meaning that according to quantum theory we can influence the world or more accurately transition from one physical state to another (and we've already seen in the relevant chapter on quantum theory that this is a very problematic assumption) - ultimately we run into the exact same problem. Results measured in a quantum experiment which is supposedly influenced by the measurer are statistical results. These results can be predicted on a probabilistic level. In other words, according to quantum theory, we can influence the world, but it will only behave according to probabilities, and once again our influence is limited by statistics, which means that it doesn't meet the definition of true freedom of choice. Without going into the details at this stage, I will demonstrate this using Schrödinger's famous cat experiment. The fact that

I'm observing the cat causes it to be in one of two states - "alive" or "dead". But while I can predict what the odds are for the cat being in any one of those states, I have no possibility of influencing this result. So, this result can't be described as an example of the influence of freedom of choice on the part of the measurer.

Split consciousness - The mind-body problem

If you've stuck around up until now, I have to let you down: I think this discussion is pointless. Ultimately, the question of free choice is based on our consciousness, in which free will develops and in which the free choice is made. The problem is that so far, no one can explain what consciousness is.

While neuroscience has developed in an incredible way over the past few decades, there is no agreed upon model which explains the phenomenon of consciousness. In the deterministic context, we've already seen the experiments first carried out by Benjamin Libet in the 1980s, repeated in variety of versions by other researchers, which showed that the sense of making a decision appears consciously several seconds(!) after the decision has begun being executed in the brain at the electric-chemical level. In other words, our supposedly conscious free choice has already begun before we even knew we were making a choice. So even in the world of neuroscience, people believe that our sense of free choice is an illusion which stems from physical or subconscious deterministic processes, which precede the conscious decision. This conclusion is also based on additional experiments which showed that a nervous stimulation of the brain has created the illusion of free choice among subjects. This result raises the claim that the neural stimulation precedes the sense of choice and so the feeling of choice stems from a material source (the brain), rather than the other way around, and so it is deterministic. Additional phenomena which reinforce the deterministic conclusion in neuroscience have been observed in cases of a split brain, that is, cases in which there's a separation between the left and right hemispheres of the brain caused by an injury or caused intentionally as part of the treatment of severe cases of epilepsy. Anyone interested in further details should read the books by Michael Gazzaniga and Antonio Damasio.

Against such conclusions from the Libet experiment and others, one can claim that there's a difference between the sense of "choice" and the sense of "deciding". These experiments show that our sense of "choice" is an illusion, but the act of "deciding" which stems from reasoning is carried out in our consciousness alone and isn't limited by neural processes. Furthermore, the existence of illusions isn't new to us and we know them from phenomena such as fata morgana or phantom pains, but these illusions aren't proof that our overall perception of reality or our overall sense of pain are merely illusions. These arguments only serve to reinforce the understanding that this discussion can never be logical or scientific, and that the answer to this question is in the eyes of the believer. I believe that even if one day we can create an entity with an artificial consciousness, which will demonstrate an ability of free choice and free will (and I'm very skeptical about this possibility), there will always be those who claim that human beings include an entity which isn't made of matter.

So, as we mentioned above, consciousness remains an unsolved riddle. Many people believe that consciousness is a product of another, non-physical entity called the mind, and is described in a variety of ways in different cultures and religions. The mind cannot be explained using science or any language which describes the physical or material universe, and anyone trying to describe the mind through these tools is like a person trying to combine words from English and Chinese into sentences, expecting them to have syntactical validity. It's impossible, and in my opinion such attempts damage the spiritual claim which loses its essence in the framework of a scientific argument which attempts to prove the existence of such an entity. Either you believe that the mind is an entity which is separate from matter and operates according to its own rules, or you believe that consciousness is an emergent product of matter, in which case you bear the responsibility for searching for the solution to the question of free choice, using scientific tools.

Anyone choosing to believe in a separate spiritual entity which is the origin of the phenomenon of consciousness, including free choice and free will, has to do deal with ancient questions regarding the relation between the body and the mind. Put simply, the problem is this: it's clear to us that consciousness influences the body, since our thoughts and feelings have an immediate expression in the material world, and it's clear to us that matter influences the mind - in cases of injury or illness, for example. How can a material entity and a spiritual entity have a mutual influence of this type, if they operate according to different rules? If they do not operate according to different rules and concepts, we must find the answers in the framework of material language, which is the scientific language, and in this language, we run into a dead end every time, at best, or a complete refutation of a spiritual component at worst. In my opinion, anyone who believes in a spiritual entity and who's interested in proving its existence mustn't stick to an approach which proves its existence scientifically. Ultimately, the spiritual component and the certainty felt by those who believe in it stem from a personal experience which is the ultimate proof for its existence. One day, we may be able to find a common language which can explain these different worlds, but such a language hasn't been invented yet. Time and time again, we find ourselves back at the psycho-physical problem, or the mind-body problem, which hasn't been solved since the dawn of human intellectual history.

If there's a separate spiritual entity, at some point it must create an electric field which will cause the electric activity we identify in the brain when a thought takes place. How is an electric field, which is a material factor, created ex nihilo? This is the question which remains unsolved. There are those who claim that the mind can do this and that's all there is to it. The limitations of science prevent us from understanding this, but it is claims like this that leave us in the realm of faith, which is perfectly fine as long as we're aware that it is a matter of faith and nothing else.

I'm not sure that any of these approaches can be proved or disproved, and we can always claim that these are two correlative states. The problem might lie in the fact that we're trying to understand a certain phenomenon of ours (consciousness) using that same phenomenon. Can we truly explain thought, when the only tool we have is thought itself? It seems like we'll never be able to do it. For this reason, libertarians always return to the claim that our intuition can't be satisfied with just material explanations and nothing else. Our thought will always have an intuitive sense that it exists in its own right, which is why this is a vicious circle argument.

In order to believe in freedom of choice, do we really have to accept mind-body dualism, i.e., the claim that that there's a spiritual entity which influences physics in a way that contradicts the laws of physics? If a certain physical neural state is associated with a certain mental state, progresses in accordance with the laws of physics to the next neural state, associated with another mental state, this is materialistic determinism and there is no room for free choice. In order for the next neural state to be determined by a choice I've made, the transition between states can't be merely physical, but rather it must be influenced by an external entity which we don't currently understand. Of course, this argument would be refuted if we find a hypothesis in physics which explains free choice and free will within the framework of neural, physical processes.

The network of choice - Free will and quantum entanglement

The question of consciousness can be made clear using network and complexity theory. Ultimately, the human brain is a network of neurons which have electric and chemical relations between them, in a highly complex structure which we don't fully understand yet. According to complexity theory, the interaction between the system's components result in emergent properties which do not exist in the components themselves. Will the understanding that the phenomenon of consciousness, which allows us to experience the feeling of free choice, is an emergent phenomenon created from the complex structure of the human brain, help us solve the question regarding our capacity for free choice? I think it will - not just because we mustn't give up on the feeling that free choice is real, but also because it's possible that network-related directions of thought regarding the structure of reality and time will pave the way for building a theory which will combine physics with our strong intuition regarding free will.

In chapter 14 I suggested that the "present" moment has a different and special status compared to "past" and "future" moments. Unlike the "block universe" view, which claims that all future moments are predetermined, according to my suggestion, the possibilities of future events are predetermined, but not all of them become realized in reality. The moments which become realized in reality are a fraction of all possible moments, and they become realized in accordance with the process of entanglement which I described, where the configuration crosses a certain threshold of connectivity which results in a "collapse" of a particular moment into reality. So, it seems that this view has a certain component of freedom given to future moments and an openness which doesn't exist in the current physical views of the block universe and the Wheeler-DeWitt formula.

Whether we're convinced that consciousness is an emergent product of the physical activity in our brain, or whether we believe it's a product of a separate spiritual component, we still need to explain the relation between physical phenomena and our consciousness, and consciousness' "free" influence on reality. I will try and make a proposal which is still in very early stages, which can possibly be developed and used to arrive at some new and interesting conclusions in this discussion.

Since every moment coming to be in reality is a result of an accumulation of quantum entanglements within a configuration and between it and other configurations, it's possible that the entanglements created between the particles or components in our brain have an effect on the configuration that ends up becoming realized in reality. This means that the fact that entanglements are created in the brain, and between the brain's components and the environment, increases the configuration's connectivity level, and promotes the realization of a certain configuration. We may be able to find a connection between the entanglements created with the environment during our conscious activity, which causes a certain configuration to become realized rather than another. In other words, it's possible that our choices create a quantum entanglement with the environment which promotes the configuration relevant to our choice over other configurations, which do not include a result of our conscious choices. If so, there exists a certain degree of freedom in the physical process according to which reality progresses from one moment to another and in which certain moments become realized while others don't, and it's possible that there's a physical connection between it and the conscious process which takes place in the brain. In all honesty this description isn't a solution to the fundamental question of what consciousness is and how a spiritual entity is created, or influences the material world, and a lot of thought must be dedicated in order to solve them.

Ultimately, consciousness is the only concept we can examine by using it, and for this reason anyone can take a stance, dig their heels in with great confidence and face contrary claims. So, the question of consciousness is in my opinion a question of faith and subjective experience and we will never be able to completely refute the claims of any side of the discussion, and for this reason it doesn't fall under the framework of scientific discussion, but rather it crosses over to other legitimate frameworks of discussion. Or in other words, to each their own.

SUMMARY:

The concept of time - more answers than questions?

"There is a Moment..."
(Dr. Mann, "Interstellar")

The journey in the search of time has found its end with a new beginning. A new start when it comes to thinking about the universe and time, based on a more in-depth understanding of the magic we live in, which leads us onwards to the next surprising station. The journey in the search of time which has accompanied the human race as long as anyone can remember will probably stay with us until the end of time. The mystery of the fundamental insights on our universe teaches us that any beam of light we manage to aim at reality reveals new, firmly locked doors to dark and interesting halls. I hope that by highlighting several new paths you weren't familiar with, I succeeded in making you think a little differently about your place in the universe, and desire to tread on those paths towards a deeper and more interesting understanding of the universe, time, and life itself.

At the end of the journey, do we find ourselves with more answers than questions? I believe it to be the case. In many ways, we're still shrouded in darkness. The documented philosophical and physical discussion which took place over the past 2,500 years has returned time and time again to profound questions regarding the existence and essence of time. While the discussion we began with accompanies us today as well to a certain extent, I hope that we succeeded in shining a light on the points of development of the human insight regarding time during this period, as well as possible avenues of progress in the future.

Parmenides and Heraclitus, who have accompanied us throughout this entire book as a dual foundation of the various world views that have popped up from time to time, represent the parallel paths along which we're treading. On one hand we have the stern Parmenides, the proponent of the eternalist view which represents a frozen world of static moments piling up, arranged one after the other like a film strip. Image after image, without any change or movement, from the dawn of time up until its end, in a fast sequence which creates the illusion of change, movement and time. It is the logical and physical time which emerges from every physical theory, every logical move in metaphysics and theological and mystical views dealing with the unity of time, as well as a divine external observer.

On the other hand, we have Heraclitus, who "flows" and "lives in the moment", the proponent of the presentist view which claims that only the present moment has any meaning or existence. Time flows from one moment to another, being created anew in each and every moment, developing freely. Nothing is known, only movement from one moment to another along the axis of time which promises us a new future, which isn't predetermined. This is our intuitive view of flow and movement in time, and of a world based on a fundamental component of constant change.

These two views have never been reconciled and used to build a unified, clear image of reality. They have collided over and over again in philosophical and physical battles, eventually ending with Parmenides getting the upper hand. The first reinforcement to the frozen pattern of the universe was supplied by Newton who presented us with a simple world which can be explained with just three laws; a world which is essentially a machine made up of different components and a known regularity which can allow Laplace to predict the future with the utmost precision. Time is just the backdrop to a predictable world progressing in a uniform rate from cause to result and the next moment which can be predicted, and the universe is made of an absolute space and time which have always existed, and which constitute the platform on which reality takes place. And then, in the last act, three figures appear in the 20th century, each of which would go on to land a death blow on the concept of time: McTaggart and his philosophical argument, Einstein and the "block universe", and Wheeler and DeWitt with the "damned" equation.

But our intuition has never given up. The sense that the physical description of time is lacking, that life must be breathed into it, has never left us. One of the main philosophers who tried to revive physical time and contend with Newton's claims was his bitter rival, Leibnitz. Leibnitz's claim, that one of the most fundamental properties of matter is a component of change, isn't an intuitive claim and hasn't stood the test of time in light Newton's claims regarding an inert matter moving along the axis of time. But today, in the second decade of the 21st century, following the "entangled configurations network" I proposed in this book, we can already say that Leibnitz's claim has a grain of truth which must be re-examined. It is very possible that this dynamic component which stems from quantum entanglement, and which is inherent to the interaction which takes place between matter's most fundamental components, is the element of change which Leibnitz hoped to find in matter itself. Like in Leibnitz's claim, the hypothesis I'm proposing also has a basic dynamic component in matter which has ontological primacy over time itself, and which is the source of time. Furthermore, Leibnitz's relativistic view of space and time, compared to Newton's absolute concepts, is much closer to the relativistic and network language of the past 100 years. Of course, Leibnitz never talked about networks, but when we take a deeper look at his view, we find in it properties which correspond to a network world view of components with relations that create the full picture. But, as we mentioned above, Leibnitz's view was sidelined during the reign of Newtonian physics for 250 years. Einstein's special theory of relativity breathed new life into the attempt to understand time, and to this day is seen by many as constituting a revolution in the view of time, compared to that of Newton. But was this an actual revolution? Einstein may have reshaped the concept of time, doing away with the absolute time which flows in the backdrop of all of reality's events, and turned it into a factor which is influenced by the matter and movement of each object. Time became a relative factor which changes in accordance with the behavior of matter in the universe, and in this sense it constitutes a revolution, but in the context of the essence of time, Einstein was in many ways a follower of Newton. He may have done away with absolute time, but when it comes to the fundamental question regarding the essence of time - whether time is a physical entity with existence in objective reality or simply an illusion resulting from other factors - Einstein simply replaced the concept of absolute time with the four-dimensional concept of time. In relativity theory, space-time is still a physical entity with existence, and all of our experience takes place within it, somehow. It's not entirely clear what space-time is, and how it's different from absolute time, save for its relative nature. It is therefore no wonder that the questions which have accompanied us since the days of Parmenides and Heraclitus have not received a significant answer following the publication of relativity theory, and for this reason I claim that in this sense, relativity theory isn't truly revolutionary when it comes to the view of time.

In fact, the 20th century has introduced a good deal of mystery to the image of reality. The relative nature of the universe's infrastructure and the magic which was gradually discovered in the fundamental components of the universe as part of quantum theory have diverted our attention to the concept of time as well, as part of the attempt to explain this mystery and magic. While quantum theory didn't directly touch on the concept of time, it managed to introduce new components to the discussion. Firstly, the amazing phenomena discovered in the particle world and the large gap between human intuition and the physical world view of the past few centuries have led us to understand that we need to dismantle existing patterns in the study of reality. Now we have the legitimization to try and understand the concept of time through new views and patterns of thought, which aren't as mechanistic and intuitive (or at least what passes for intuitive for a scientist at the end of the 19th century or the start of the 20th century). Quantum theory has also provided us with new tools with which to better understand not just the behavior of the basic components which make up the universe, but also the macroscopic components and the infrastructure of space and time.

Ultimately, despite countless attempts to find the ultimate definition for the concept of time, one which could meet both the standards of physics which described a static universe, and those of the intuitive insights of a flowing, renewing time, we've arrived at the 21st century with two opposed views: the death of time and its rebirth according to Barbour and Smolin, respectively, without a solution which can bridge between the two. 2,500 years have passed and Parmenides and Heraclitus' ideas have been developed and refined, as well as being better founded both physically and philosophically, but they are the same old things under a different cloak.

The age we live in brings with it a new scientific revolution. The discoveries of the past thirty years have decisively transitioned us from mechanistic views of reality to organic ones, and from the attempt to understand the universe by studying its components to a view which studies the system as a whole, as a complex entity which operates in its own right, creating new processes and phenomena. Network and complexity theory may have focused so far on the fields of biology, computers and sociology, but after a long period of thought and study, we can already use the insights which followed from the study of these worlds and try and apply them in the worlds of physics and metaphysics in general, and in the attempt to understand the concept of time in particular. These insights have clarified how a complex system develops and how new and fascinating emergent phenomena are created, ex nihilo, from the interaction of components which make up the system. They have significantly improved our ability to understand and explain complex and mysterious phenomena in the world and have even provided us with a different and interesting point of view of the way in which each of us perceives their personal, day-to-day reality - from our relationships with the people around us to building a career and managing a business; they can even serve as a basis for constructing a life philosophy which can deal with questions of happiness and personal success.

One of my main goals in writing this book was to try and combine all of these components provided to us by the 20th century, within the framework of quantum theory and complexity theory, with new insights and hypotheses which arise from the physical world as we speak, into one overall hypothesis which could try and tackle the question of time.

As part of this move, I described the "entangled configurations" hypothesis at the end of this book, which in my opinion has great potential to open a new path of thought and study in these worlds. The "entangled configurations network" is essentially a fundamental hypothesis on the structure of the universe as a collection of moments which include arrangements (configurations) of matter in the universe, in accordance with the physical view which follows from the "block universe" hypothesis and the insights which are elegantly summarized in Julian Barbour's book. However, unlike my predecessors, I don't mean to leave this collection of moments as a pile of static time, collecting dust, unchanging, and without the possibility of choice between different paths of progression. My goal is to breathe life into these frozen moments which will be able to explain our intuitive experience of the universe as a living, flowing universe, which progresses from one moment to another in a process of renewal and becoming which isn't predetermined. For this purpose, I made use of the fascinating phenomena which have entered our lives following quantum theory, in the form of quantum entanglement and superposition, and the tool box provided to us by network and complexity theory. The result is a universe whose foundation is a network of "moments" or arrangements of the matter in the universe, in a certain state of existence which hasn't been realized in reality yet (superposition). Once I assumed the existence of the network and allowed it to develop on its own, I discovered to my amazement that the network's development and the interaction between its components, in accordance with the regularity we've encountered in network and complexity theory and from the connectivity based on the phenomenon of quantum entanglement, create surprising and fascinating emergent phenomena, led by the realization of all the nodes which make up the network, one after the other, as moments of time in physical reality. All of a sudden, the "block universe" and Barbour's pile of static "moments" have received life due to a basic component of change in the universe (quantum entanglement) which allows for connectivity and synchronization, whose existence has primacy over the moments of time which become realized in reality. Entanglement is the basic component which Leibnitz sought in the essence of matter; entanglement, which is a necessary condition for the combination of the various contradictory views of the concept of time and the completion of the image of our intuitive image of reality.

In the beginning, the earth was formless and empty, and from the chaos an order was created which was renewed in every moment, time and time again, from the chaos: the fireflies' light show along the rivers of south east Asia, the awe-inspiring choreography of a flock of birds in the autumn sky, the destructive results of global financial crises, human society which is reshaped in the framework of social networks and the mysterious consciousness which breaks through a network of neurons. And first and foremost, a universe created from one moment to another, from the "network of time".

***

On a personal note, I have spent many years dreaming about the moment in which this book gets published, helping you to better understand reality and the universe in which we live, to shatter the patterns on which we were raised and to rethink everything that seems obvious to us in the universe and our personal lives, and mainly on the basic component which constitutes the infrastructure of everything that we are - the concept of time. Every moment in our lives is realized from an infinite collection of possible moments. We don't know if we have a choice when it comes to the next moment, but in any case, it is unique, and a result of the magic called "the universe". All I can do at this point is thank you for those moments from your collection of moments which you chose to devote to making my dream come true.

BIBLIOGRAPHY

- Abott Edwin A., Flatland: A Romance of Many Dimensions, Dover Publications (1992)
- Barabási, Albert-László. Linked: The New Science of Networks, Basic Books (2014).
- Barbour, Julian. The End of Time: The Next Revolution in Physics (New York: Oxford University Press, 1999).
- Bardon, Adrian. A Brief History of the Philosophy of Time. New York: Oxford University Press, 2013.
- Canales, Jimena. The Physicist and the Philosopher: Einstein, Bergson and the Debate That Changed our Understanding of Time (Princeton: Princeton University Press, 2015).
- Capra, Fritjof. The Tao of Physics (Boston: Shambhala Publications, 1991).
- Carrol, Sean. From Eternity to Here - The Quest for the Ultimate Theory of Time, Dutton (2010).
- Damasio, Antonio. Descarte's Error: Emotion, Reason and the Human Brain (Penguin, 2005).
- Desmurget, Michel, Reilly, Karen T., Richard, Nathalie, Szathmari, Alexandri, Mottolese, Carmine and Sirigu, Angela, Movement Intention After Parietal Cortex Stimulation in Humans, Science 324 (2009): 811–813.
- Duem, Pierre. The Aim and Structure of Physical Theory (Princeton: Princeton University, 1991).
- Evans, Denis J. and Searles Debra J. The Fluctuation Theorem, Advances in Physics 51. (2002): 1529–1585.

- Fleck, Ludwik. Genesis and Development of a Scientific Fact (Chicago: University of Chicago Press, 1979).
- Gazzaniga, Michael. Who's in Charge? Free Will and the Science of the Brain (New York: Ecco, 2012).
- Genovese, Marco. Time from Quantum Entanglement: An Experimental Illustration, Physical Review A 89 (2013).
- Gimble, Steven. Redefinig Reality: The Intellectual Implications of Modern Science. The Great Courses, 2015.
- Gleick, James. Chaos: Making a New Science, Penguin Books (2008).
- Green, Brian. The Fabric of the Universe: Time and the Fabric of Reality, Knopf (2020).
- Grim, Patrick. Philosophy of Mind: Brains, Consciousness, and Thinking Machines, The Great Courses, 2013 (audio recording).
- Heinz R. The Dreams of Reason: The Computer and the Rise of the Sciences of Complexity (New York: Bantam, 1989).
- Johnson, Neil. Simply Complexity: A Clear Guide to Complexity Theory. London: Oneworld Publications, 2007.
- Kauffman, Stuart. Investigations (New York: Oxford University Press, 2000).
- Kauffman, Stuart. The Origins of Order: Self-Organization and Selectionin Evolution. New York: Oxford University Press, 1993.
- Kuhn, Thomas, The Structure of Scientific Revolutions (University of Chicago Press, 1996).
- Loewer, Barry. Freedom from Physics: Quantum Mechanics and Free Will, Philosophical Topics 24 (1996): 91–112.
- MacTaggart, J.M.E. "The Unreality of Time". Mind 17, 1908, p. 457–473.

- Maldacena, Juan. Entanglement and the Geometry of Spacetime, (2103, https://www.ias.edu/about/publications/ias-letter/articles/2013-fall/maldacena-entanglement).
- Margenau, Henry. Quantum Mechanics, Free Will and Determinism, The Journal of Philosophy 64 (1967): 714–725.
- Melanie Mitchell, Complexity: A Guided Tour (Oxford: Oxford University Press (2009).
- Penrose, Roger. The Emperor's New Mind: Concerning Computers, Minds, and the Laws of Physics (New York: Oxford University Press, 1989).
- Popper, Karl. The Logic of Scientific Discovery. New York: Routledge, 1959
- Price, Huw. Time's Arrow and Archimedes' Point: New Directions for the Physics of Time, Oxford University Press; New edition (1997).
- Prigogine, Ilya and Stengers, Isabelle, Order out of Chaos: Man's new Dialogue with Nature (New York: Bantam, 1984).
- Rovelli, Carlo. Reality Is Not What It Seems: The Journey to Quantum Gravity. New York: Riverhead Books, 2016.
- Sagan, Carl, Cosmos (New York: Random house, 2002)
- Smolin, Lee and Mangabeira, Roberto Unger. The Singular Universe and the Reality of Time (New York: Cambridge University Press, 2015).
- Smolin, Lee. Time Reborn: From the Crisis of Physics to the Future of the Universe (New York: Houghton Mifflin Harcourt, 2013).
- Strogatz, Steven. Sync: How Order Emerges from Chaos in the Universe, Nature and Daily Life (New York: Hyperion, 2004).
- Susskind, Leonard. Quantum Complexity Inside Black Holes" (A Lecture at Stanford University, 23/10/2014,

https://www.youtube.com/watch?v=FpSriHE1r4E).

- Susskind, Leonard. ER=EPR But Entanglement Is Not Enough (A Lecture at Stanford University, 4/6/2015, https://www.youtube.com/watch?v=IuY4RMehdP8).
- Susskind, Leonard and Maldacena, Juan. Cool Horizons for Entangled Black Holes (2013, http://arxiv.org/abs/1306.0533).
- Swingle, Brian. Constructing Holographic Spacetimes Using Entanglement Renormalization (2012, http://arxiv.org/abs/1209.3304).
- Tetlock, Philip E. and Dan Gardner. Superforecasting: The Art and Science of
- Prediction. New York: Crown, 2015.
- Van Raamsdonk, Mark. Building up Spacetime with Quantum Entanglement, General Relativity and Gravitation 42, (2010): 2323–2329.
- Vedral, Vlatko. Decoding Reality: The Universe as Quantum Information. New York: Oxford University Press, 2012.
- Waugh, Alexander. Time: From Micro-Seconds to Millennia – A Search for the Right Time. London: Headline, 1999.
- articles published by Prof. Yakir Aharonov appear in his website: http://www.tau.ac.il/~yakir/, in particular, see article 169 on the site, on the subject of "weak measurements".

Made in the USA
Columbia, SC
05 September 2023